Mass Customization and Footwear: Myth, Salvation or Reality?

Claudio R. Boër and Sergio Dulio

Mass Customization and Footwear: Myth, Salvation or Reality?

A Comprehensive Analysis of the Adoption of the Mass Customization Paradigm in Footwear, from the Perspective of the EUROShoE (Extended User Oriented Shoe Enterprise) Research Project

Springer

Claudio R. Boër, Prof., PhD
Director
ICIMSI–Institute of Computer Integrated Manufacturing for Sustainable Innovation
SUPSI–University of Applied Science of Southern Switzerland
Galleria 2
6928 Manno
Switzerland

Sergio Dulio, Eng.
Via Simone del Pozzo 18
27029 Vigevano
Italy

British Library Cataloguing in Publication Data
Boer, C. R. (Claudio R.)
Mass customization and footwear : myth, salvation or reality? : a comprehensive analysis of the adoption of the mass customization paradigm in footwear, from the perspective of the EUROShoE (Extended User Oriented Shoe Enterprise) Research Project
1. Footwear industry 2. Mass customization
I. Title II. Dulio, Sergio
338.4'76853

ISBN 978-1-84996-662-7 e-ISBN 978-1-84628-865-4 Printed on acid-free paper

9 8 7 6 5 4 3 2 1

Springer Science+Business Media
springer.com

Preface

Mass, customization and footwear: these three words appear in the title of this book and summarize its contents and purposes. The word *mass* reminds us of the many individuals that make up a community, the population for whom consumer products are normally thought. It also reminds us of the traditional organization of industrial production, of the way large quantities of almost identical items have been produced until today. It recalls the past. *Customization* deals with individuality, with personal tastes and with consumer satisfaction: the challenge for tomorrow is to offer the privilge of individuality to the masses rather than just to elite consumers. A big challenge, which is even bigger when the goods to be produced are *footwear*, with all the peculiarity of a complex product, with a complex industrial pipeline in a market dominated by laws that, until today, seem to privilege big numbers rather than niches. This is the aim of this book: to show how mass, customization and footwear can live together in the industrial scenario of the years to come and, finally, how mass customization can contribute to local and global sustainability.

Claudio R. Boër
Sergio Dulio

Acknowledgments

The authors would like to thank the many people whose work has made this book possible. It will be difficult not to forget somebody but we have tried our best.

First, we want to thank the European Commission Research Directorate who funded the project EUROShoE within the 5th Framework Program, and in particular Mrs. Roberta Salonna, Project Officer, who gave critical but always constructive contributions through the whole project. Without her support it is certain that the project would have had much more difficulty in reaching its goals.

Second we would like to thank all the personnel of ITIA-CNR who, in one way or another, have helped during the whole project, from its start to its conclusion. In particular we would acknowledge the invaluable contribution of Eleonora Schiariti who was not only a secretary but also a very attentive organizer of meetings, documentation and project events.

Andrea Zangiacomi was an important gear in the EUROShoE project and his contribution was not only as a researcher but also as a very competent and precious financial administrator of a complex consortium involving so many partners.

Claudia Redaelli was not only a researcher but also a great support in collecting information and in interpreting them; her contribution is particular evident in the chapter regarding customer satisfaction and the requirements for footwear customization.

Rosanna Fornasiero is also acknowledged for her important contribution in Chapter 2 where a model for analyzing the financial aspects of mass customization in the footwear industry is presented. The value of this model is not only theoretical but it is also applicable in practice; by adopting the correct parameters and data, companies interested in entering the business of mass customization can simulate their own environment to select the best approach.

Furthermore, we would like to thank Cristoph Berger (Adidas) and Claudia Kieserling (Selve), who have provided data and information about their approach to mass customization, and Frank Piller, whose profound knowledge on mass customization and personalization has pushed us always forward in our project, indicating new actors, new approaches, and new sectors in which mass customization is more and more applied.

Finally we would like to mention here all the partners of the EUROShoE project:

Itia – Cnr (Italy)
Bally Schuhfabriken (Switzerland)
Calzaturificio Frau (Italy)
Calzados Anatomicos Calana (Spain)
Jefar Industria De Calcado (Portugal)
Lirel – Lima & Resenda (Portugal)
Lloyd Schuhfabrik Meyer &Co (Germany)
Ecco Sko (Denmark)
Fagus-Grekon Greten (Gremany)
Formificio Milanese Team (Italy)
Consorzio Sintesi (Italy)
Centro Servizi Calzaturiero (Italy)
Csm3d International (United Kingdom)
Graisoft (France)
Massen Machine Vision Systems (Germany)
String (Italy)
Atom (Italy)
Comelz (Italy)
Molina & Bianchi (Italy)
Torielli Rag. Pietro & C (Italy)
PFI Pruf Und Forschunginstitut Fur Die Schuhherste (Germany)
Siemens (Italy)
Delcam (United Kingdom)
Technische Universitaet Muenchen (Germany)
Fondazione Don Carlo Gnocchi – Onlus (Italy)
National University Of Ireland – Galway (Ireland)
Institute CIMSI of SUPSI (Switzerland)
Fraunhofer Gesellschaft Zur Forderung (Germany)
Asociacion Instituto De Biomecanica De Valencia (Spain)
Universitaet Karlsruhe (Germany)
Universitaet Hannover (Germany)
Loughborough University (United Kingdom)
University of Genoa (Italy)
Joseph Stephan Institute – Lubljana (Slovenia)

With them we shared around four years of knowledge and experience in a challenging technological adventure.

Contents

1

Introduction: Why this Book and to Whom it is Addressed

This book is about mass customization but, perhaps more than that, it is about footwear; we think it is important to stress this point in order to clarify the specific perspective that we decided to adopt in treating the main topic of the book. There are already several books and a large number of scientific and popular publications about the concepts of product personalization and mass customization. But we are not aware of any other in depth treatment of the peculiarities of the application of this business model and manufacturing paradigm to a specific industrial sector, in our case footwear.

The purpose of our work was to provide a comprehensive presentation of the many possible different applications of mass customization to the shoe as a product and to shoemaking as an industry. The book is also about the project EUROShoE, which, for the first time, brought together a large number of stakeholders of the footwear world in a cooperative research project that focused on the application of mass customization to footwear. As an outcome of their joint effort, models, enabling technologies and a pilot implementation were developed that have shown shoe makers that product customization was possible and that it could be considered as a valid, workable option to improve their competitiveness. In writing these notes today, we can certainly state that the general acceptance of the concept among shoe companies was much higher at the end of the project than it was at its beginning; and it is even higher now, more than two years after the conclusion of the research.

We did not intend to write another textbook on the general concepts of mass customization, so readers will not find anything more than some basic standard definitions on that issue [1]. They will, on the other hand, find a quite detailed analysis on how the ideas of mass customization apply to footwear, which problems are to be faced in implementing it, and the factors to which most attention should be paid. A methodology for calculating the return on investment based on different levels of implementation of mass customization is also presented, because in a competitive industry it is not enough to speak about new ideas and enabling technologies if the basic concepts cannot be validated by the company cheque-book. This section of the book sets the framework for the correct implementation of mass customization in footwear companies.

Technology plays an important role in enabling the adoption of the new business model at all levels in companies, allowing the modification of all relevant processes to become customer orientated. Hence a central role in the book is taken by the presentation of the research project that first studied the kind of technologies that were needed, defined their specifications and produced test bed implementations of most of them. Many of these technologies have now been developed in industrial applications used for mass customization.

We also believed that the best way to prove that the idea was realistic, and that many different implementation paths were possible, was by showing examples and best practice cases; therefore the central part of the book takes an in depth look at different ways mass customization can be adopted by shoe companies, for each one of these ways, examples and detailed case histories are given of those companies that have already successfully moved along that path. Finally, in the final section of the book, we try to draw some final conclusions on whether the knowledge developed and the lessons learnt on the topic of mass customization and footwear could be of any interest and application to other industrial sectors.

Is mass customization a strong enough new competitive factor to help the footwear companies of the most industrialized countries in facing the current difficult economic situation? Is it a possible salvation from the fierce competition on quality and price they have to fight in most of their markets? Although no final conclusions are possible yet, we have nevertheless tried to give a few indications on what is the situation now in regards to the penetration of the idea of mass customization, on its potential for the future and on what is likely to happen in the years to come.

If we were asked why we decided to write this book, probably the easiest answer would have been: to share with those involved in shoes and shoemakers the knowledge that we and the other members of the team who supported us in preparing this book have gathered on the topic of mass customization applied to footwear. Our aim has been to provide shoe practitioners with examples to follow and basic guidelines to help them evaluate the feasibility of this approach for their common business. Practitioners as well as curious readers fascinated, on the one side, by shoes and their world and, on the other, by the idea of linking that with a modern interpretation of consumer centricity, are the public we had in mind in preparing these notes.

We also think that the information, the concepts and the ideas that are presented in this book could be interesting for graduate students studying in the area of business management as well as in the area of design and manufacturing engineering, where a strong interaction between the two disciplines, we believe, is necessary in order to have a successful implementation of the presented concepts. In addition, the book may be of some interest to researchers and academics who are looking for an in depth presentation of an industry specific application of an interesting emerging paradigm as mass customization is. With these goals in mind we now submit this endeavour to the attention of the readers with the hope that they enjoy reading it as much as we have enjoyed the experience of writing it.

2

Mass Customization and Footwear

2.1 The Mass Customization Paradigm and its Production Revolution

Manufacturing has been the prime driver in the evolution of society from one that is agriculturally centred to one that is industrially centred. However, manufacturing has also evolved through time, and several paradigms can be identified as described in Table 2.1. [2]. One of the main trends in today's market is that of 'mass customization'. This represents a new market paradigm that is changing the way consumer products are designed, manufactured, delivered and recycled.

Manufacturing technology started with an artisan at work making a single product for a single customer, and as such was well recognized as craft production, as illustrated in Fig. 2.1.

Manufacturing continued to evolve in the late 1800s during the Industrial Revolution, pioneering mass production at the beginning of the twentieth century. Today this market of mass production is changing and moving towards the new paradigm of mass customization. It is thus recognized that the current and future manufacturing challenges are returning to those of the original craft production age, but with the added advantages and complexities of using today's advanced manufacturing systems and technologies. Therefore, one view of mass customization could be as having the ideals of craft production expressed through modern industrial technology.

As identified by [3], mass customization aims to offer goods and services that are more tailored to customers specific needs and tastes, which implies having the capability of linking the efficiency and economy of scale of mass production with the possibility of manufacturing small batches and batches of one of very diversified and personalized products.

Table 2.1. Evolution of production paradigms

Paradigm	Craft production	Mass production	Flexible production	Mass customization	Sustainable production
Started	~1850	1913	~1980	2000	2020?
Society needs	Customized products	Low cost products	Variety of products	Customized product	Clean products
Market	Very small volume per product	Demand > supply Steady demand	Supply> demand Smaller volume per product	Globalization Fluctuating demand	Environment
Business model	Pull *sell-design-make-assemble*	Push *design-make-assemble-sell*	Push-Pull *design-make-sell-assemble*	Pull *Design-sell-make-assemble*	Pull *Design for environment-sell-make-assemble*
Technology enabler	Electricity	Interchangeable parts	Computers	Information technology	Nano/bio/material Technology
Process enabler	Machine tools	Moving assembly line	FMS[1] robots	RMS[2]	Increasing manufacturing

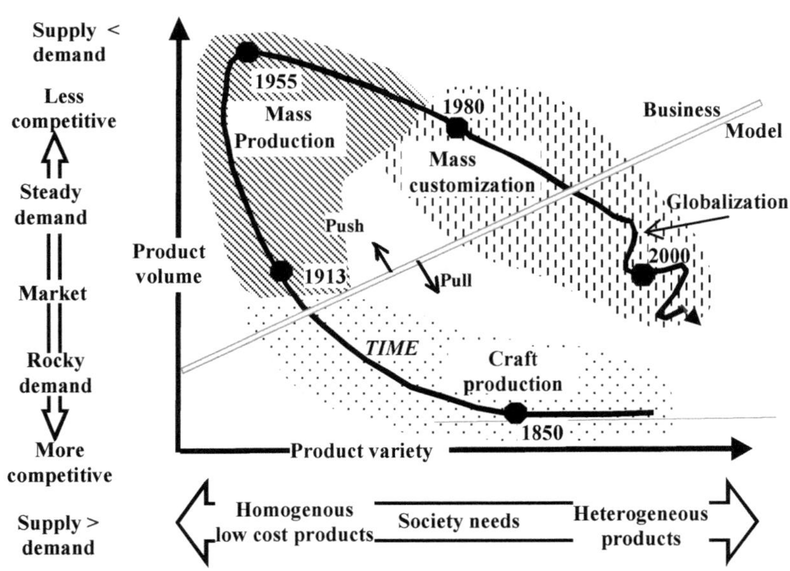

Figure 2.1. Evolution of manufacturing technology

[1] FMS = Flexible Manufacturing System
[2] RMS = Reconfigurable Manufacturing System

This places a very strenuous challenge to the entire company organization, whose procedures and management approaches then require a thorough revision, especially in manufacturing. This is certainly true for shoe production; as footwear manufacturing is increasingly confronted with a progressive reduction in the size of production batches. Combined with the variability of styles, this tends to overstretch the traditional work organization and, with a demand for minimizing delivery times, manufacturing support systems do not as yet approach the levels of flexibility and quick response required for the production of mass customized products. However, since a noticeable demand for such products is becoming evident among shoe consumers, footwear companies will soon have to confront these kinds of technical challenges.

2.2 The Footwear Business at the Start of a New Millennium

Making shoes is neither an easy task neither a simple business. It necessitates high skill and a lot of diverse knowledge in many aspects that may affect the quality, the aesthetics and the functions of a shoe, but, more importantly, it requires a lot of work, which makes shoemaking a typical labour-intensive activity. Shoemakers of the past, as well as those who are still continuing this tradition today, know very well what this means in terms of the time needed to produce a pair of shoes and the complexity of handling the manufacturing and assembling of all its different parts.

More than a century ago, when the first machines appeared and shoemaking evolved from a craftsman activity into an industrial one, large numbers of workers were needed to produce the higher quantities of shoes that the industrial production made possible and that an expanding market required, in particular, in the economically developed countries. With time, with the advent of better work organizations and of more modern machines, the situation improved. A typical "indicator" of the importance of the labour force was the increase in the average number of pairs produced daily by each direct worker, yet at a much slower pace than in other industries that were undergoing similar transformations in the same times.

During the last 40 years, with the widespread adoption of information and communication technologies, computers and process automation, progress was certainly made and shoemaking did in fact modernize, particularly in terms of quality of the manufactured products, flexibility of production, level of control on the various processes, consistency and constancy of the quality of the delivered products and so on. Yet it still requires a large number of workers to achieve an acceptable production throughput that can support industrial scale operations. Perhaps this kind of statement is less of a general nature than a few decades ago; certain kinds of shoe typologies and their respective constructions are more easily performed than others (like, for instance, making a pair of sneakers in comparison to producing high quality Goodyear welted shoes), they require a smaller number of manufacturing steps and they lend themselves to an improved usage of automated machines and of unattended processes. These kinds of shoes, either for their high quality or for the high level of automation in their manufacturing process, are among those that are still produced in areas like Western Europe and

Latin and North America; moreover, labour demands concentrate in some of the phases in which a typical shoe manufacturing process is structured (such as the so-called cutting and upper stitching phases), which has generated another typical shoe making phenomenon: subcontracting and delocalization of the most labour intensive phases of the production process.

It is important to understand why and how the footwear business has changed and evolved in the last 50 years. Labour, we have learnt, is very relevant to shoe production (both in terms of the number of workers and of the skill they need to have to perform their task), and, consequently, labour cost is certainly a very relevant factor in determining the total manufacturing cost of a shoe and, eventually, in fixing the final price. So shoe makers have always struggled to keep their labour costs to a minimum.

This is what has triggered the subcontracting and delocalization phenomenon of shoe producers, who became more exposed to endogen factors (more attentive consumers always asking for a higher quality at a competitive price) and exogen factors (competitors from other countries which were capable of delivering similar products at a better price), and therefore had to look for means of keeping their competitive position; moving the most labour-intensive phases of production to lower wage countries seemed, to shoemakers as to other goods producers, the easiest and faster solution.

While there is no doubt that this was the easy way out to what was becoming an impellent survival problem, certainly this did not turn out to be, strategically thinking, the most clever move one could have thought of. This has generated a "caravanning" effect with the big shoe companies of the more economically developed countries (first of all the United States) progressively moving their production facilities to third world or developing countries, where salaries and wages were lower and where an abundant workforce was available; as wages started to grow in those countries too and the workforce became less available, these companies then had to relocate to other countries which were, at the time, the low wage champions. This phenomenon is typical of products whose content involves less innovation. It can be avoided through attention to such things as the demands of the end consumer, with innovative materials, or modern processes; all factors that can reduce the labour component of the product price.

With time, rather than some of the processing steps, the entire manufacturing process was subcontracted; and that implied installing in the target countries fully fledged manufacturing facilities, transferring technologies, work methodologies, knowhow and skill. In a few decades, in those countries where resources (both natural and human) were readily available, students became better than their teachers and started to compete with them in international markets. Or, worse, with other shoemaking countries that had no direct responsibility in this transformation. Perhaps other factors have played a role, as it is in the nature of the economy and of industrial evolution that labour-intensive manufacture is typical of developing countries with an abundant workforce and for which work is a "social asset" (rather than a cost to be controlled) but bound to disappear from those that are economically developed. Nevertheless, these factors are to be kept in mind when analyzing shoe production today.

In order to understand all this, we have to look at numbers, and numbers, of course, change with time. So, rather than considering absolute numbers, it is better to focus attention on the evolution, *i.e.* on how numbers change with time and the underlying trends. There are many figures to support this analysis, but two that are particularly relevant to the footwear world are:

- the world yearly consumption of shoes
- the world footwear production

Finding reliable data such as the world yearly consumption of shoes is not easy. The most recent data presented in 2005 indicate an annual world consumption of shoes (in 2003) of roughly 16.9 billion pairs.[3] During the period from 1993 to 2003 the level of consumption grew at an average annual rate of 3.21%, with sustained growth in areas such Asia (5.0%), the Middle East (3.25%) and North America (2.09%). In other areas of the world growth was modest (*e.g.* Western Europe, at 0.36%). As we will see later, the analysis of volumes is only one aspect of consumption - values will also need to be considered.

One interesting analysis would be to relate annual consumption of shoes to total world population; this indicator would tell us how many pairs of shoes are consumed per person per year. If data on footwear consumption are difficult to assess, then statistics on the world population are even more difficult to obtain. Nevertheless, taking the US Census Bureau data[4] as our point of reference, we can calculate the number of pairs of shoes consumed per person in 2003. This suggests a value of 2.7. The same calculation for 1993 indicates a value of 2.2 pairs of shoes per person per year. In other words, in ten years the world population grew by something like 20%, whilst the percapita consumption of shoes increased by almost 23%,[5] which seems to indicate a rate of growth in consumption only marginally higher than the growth of the population.

What this analysis tells us is that growth in shoe consumption is sustained by the growth of the world population, and that such growth is concentrated in developing countries. Consumption is growing in countries where two factors are at work:

- population growth
- a developing economy with an increase in the size of social segments whose purchasing power covers necessities and also allows for the purchase of several pairs of shoes each year.

The United States appears to be an exception. Whilst the USA does not have a fast-growing population, it nevertheless shows high growth in shoe purchases. There may be two reasons for this:

---

[3] Data for 2003 presented at the second World Footwear Congress (WFC), held in Brussels in April 2005.

[4] Total world population in 2003 according to the US Census Bureau, updated 26 April 2005, was 6,303 billion.

[5] Again based on shoe consumptions data presented at the World Footwear Congress of April 2005.

- America is devoted to consumption
- a high level of immigration, particularly from Latin America, is present.

In regions such as Western Europe, where there is no population growth and consumption is modest, there was only a small increase in shoe purchases in the decade 1993–2003. To summarize, the demand for shoes grows every year at a relatively slow rate, mostly sustained by developing economies with a growing population. All this calls for a mature market, where footwear companies find it more difficult to establish and maintain a competitive advantage.

Table 2.2. Top ten consuming regions of the world (2003) (Source: WFC 2005)

Region	Q_{2003}	$Share_{2003}$	$Growth_{1993-2003}$	**Δ 03/93**
Western Europe	1,841	10.87%	0.36%	108
Eastern Europe	960	5.66%	0.85%	-3
Middle East	757	4.47%	3.25%	248
Africa	786	4.64%	1.07%	136
Asia Pacific	9,326	55.05%	5.00%	3,731
North America	2,429	14.34%	2.09%	390
Latin America	842	4.97%	0.93%	73
World	16,940	100.00%	3.21%	4,683

Regional shoe consumption has changed noticeably in the decade 1993–2003: Table 2.2 shows the top ten consuming regions in the world and the share each one of them took of annual shoe production (in 2003), as well as the evolution of such consumption in the decade 1993–2003.

With such a shoe consumption scenario, it has to be expected that production will run parallel. In fact, statistics indicate that, for the year 2003, there was a total world production of almost 17 billion pairs of shoes, with an annual growth rate, over the decade 1993–2003, of 3.34%. That enabled production to keep up with the increase in consumption and in the level of demand that we have discussed so far.

Asia takes 72.3% of this world production with an average annual growth of 4.9%; its share of the world production has increased at an annual rate of 1.57% in the same period. Western Europe represents no more than 4.58% of world production and its share over the total has decreased at a rate of 7.5% over the decade. A more impressive loss than the already noticeable Asian growth rate (Europe's share sank more than the Asian share soared).

Once more what is noteworthy of the current scenario is how much this picture has changed over time; if we go back another few years with respect to 1993, we can see in Fig.2.2 a breakdown of world production in 1989 and notice the predominance of Asian production-accounting for about 55% of the total (no disaggregated figures for China are available for that year). In addition, two aspects are worth mentioning:

- Europe (Western Europe and Eastern Europe and CSI) represented 28% of the world shoe production. This was the year of the dissolution of the USSR when a lot of big state-owned shoe factories still existed in the Soviet regions producing millions of pairs of shoes every year.
- Central and North America (which meant primarily the United States) accounted for 7% of the world production, equivalent to a few hundred million pairs of shoes in that year.

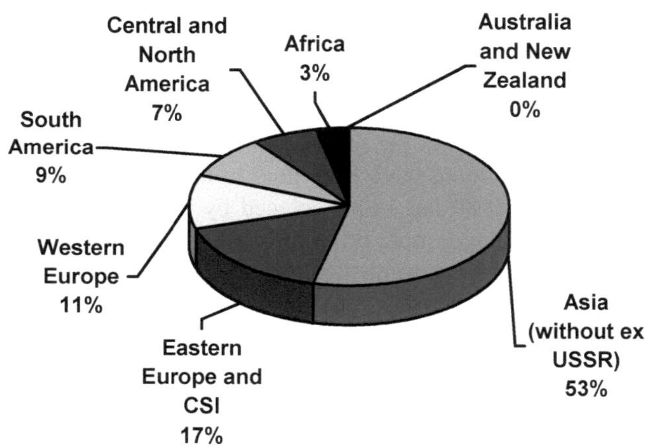

Figure 2.2. Shares of the annual shoe production in year 1989 (Source: SATRA)

In 1989, world shoe production polarized around two major manufacturing blocks: the Far East and the West (including Europe and the Americas), with the first predominant. There was open competition at that time, and the two blocks appeared to have equal chances. China was already a major contributor to the volume of Asian production, but had not yet moved into the spotlight. The reason is probably that China's exports of shoes were not such as to provoke the cries of alarm seen more recently.

Over the 14 years to 2003 (the most recent year for which consolidated figures are available), the picture has changed dramatically: Western and Eastern European countries (the latter mainly countries of the former Soviet block now on the verge of joining the European Union that have been capable of maintaining a relatively solid footwear industry[6]) now cover no more than 6.7% of world shoe production (in absolute terms approximately 1.1 billion pairs of shoes); North

[6] Some of these countries deserve a special comment; for instance, Romania, which in the last few years has become a decentralized manufacturing district of Italy, following moves by dozens of Italian shoe makers to place almost their entire production in the country.

America dropped to 1.9% of world production, having lost almost all of its manufacturing infrastructures and having dispersed its shoe making knowhow (with the exception of a group of companies maintaining manufacturing facilities in the United States).

The region that emerges as the declared winner of this world production battle is certainly Asia and the Pacific, which accounted for almost 80% of global shoe production in 2003. Two Asian tigers are leading the game: China is credited with 43.6% of world shoe production (7.4 billion pair of shoes made in 2003), followed by India with 22.1% (3.7 billion). To defend the performance of Western shoe producers, Brazil comes third with 3.59% of the world total, but next is Vietnam which, in less than ten years has been able to grow from an almost non-existent level of production to 3.24% (half a billion pairs of shoes in 2003), and then Indonesia (3.1%). We have to reach the sixth position to find the first European country, Italy with 2.15% of world shoe production (although an update of these figures based on the latest statistics for 2005 show a more negative picture). The complete list of the top ten shoe producing countries is presented in Table 2.3.

This table provides a basis for an analysis of the footwear business that is different to the one that would have been able to make only 15 years ago, where the bipolar situation of 1989 has been replaced by a more complex, three centre picture that is correspondingly more difficult to explain and to interpret. In order to perform such an analysis, we must also consider how import and export flows have been established in the past years.

In the last year of the decade considered, almost 7.1 billion pairs of shoes were exported from producing countries. Exports increased over ten years at an average rate of 4.37%. Regions other than Latin America have experienced a positive growth rate. Asian exports account for almost three quarters (75.4%) of total world exports, with an annual increase in excess of 5%. Western Europe is the second exporting region, but its annual growth rate in the decade analysed only scored a very negligible 0.05%, which caused, not surprisingly, a loss of market share of 4.32% over the same period.

The leading exporting countries are, again, a small group of Asian nations (China, Vietnam, Indonesia and Thailand). Among Western European countries, Italy and Spain are the most relevant exporters. But whilst some of the top four featured notable growth rates in the past decade, others, such as Italy, showed a decrease of around 2% per annum. It can be concluded that the big shoe producers are also big exporters, absorbing an ever-increasing quota of world footwear exports. Western countries are also exporters but their overall share is decreasing. These countries are rapidly loosing their competitive advantage in favour of big producers and of those with low labour costs, with an evident shrinking of both absolute production and export volumes.

Quantities (in this case, numbers of pairs) are normally used to measure and analyze aspects such as production or export flows; they are directly related to the production throughput of a certain manufacturing system. The higher the volumes, the higher the size and the number of the factories and, hence, the number of workers employed. It is an indicator that can show how much an industrial sector can contribute to the global workforce of a nation. But it is not in itself a sufficient indicator of how much the same production contributes, for example, to the national gross product or to the net trade balance of the same sector. For these kinds of analysis, the value of production is a second very important parameter to be considered.

Table 2.3. Top ten shoe producing countries (Source: WFC 2005)

Country	Q_{2003}	**Share**$_{2003}$	**Growth**$_{1993\text{-}2003}$	**Δ 03/93**
World	16,972	100.00%	3.34%	4,739
China	7,400	43.60%	7.31%	3,800
India	3,750	22.10%	4.45%	1,300
Brazil	610	3.59%	1.02%	27
Vietnam	550	3.24%	13.44%	415
Indonesia	520	3.06%	1.22%	180
Italy	365	2.15%	-3.71%	-162
Nigeria	355	2.09%	4.12%	120
Thailand	280	1.65%	-3.35%	-30
Turkey	280	1.65%	4.92%	125
Pakistan	250	1.47%	2.29%	65

It is then interesting to note that, if world export figures are analysed in terms of values (millions of US$) rather than volumes (billions of pairs), we obtain a slightly different picture. China still keeps its prime role with 4.5 billion pair of shoes exported (equal to 63% of world exports), which in 2003 where 12,955 million worth of US$ (only 25.7% of the world total in value); but Italy, which was third in the world top ten exporting countries (305 million pairs of shoes, equal to 4.3%), now jumps to the second place with export worth 8,479 million US$, with a more relevant share of the global export pie (16.8%). So, values make a difference; not such that it can compensate for the enormous production and export gaps between Asia and other regions, but big enough for the footwear sector to contribute to the GNP of the leading producing countries. By comparing volumes and values of exports, one can obtain very interesting indications on the average "export price" (which in turn is strictly related to the manufacturing costs) per pair of the shoes coming from the different world regions. It can be concluded that Chinese shoes are exported at an average price per pair of US$ 2.9, while in the second country (Italy) the price would be US$ 27.8. Very noticeably, we are considering quite different kinds of shoes. Vietnam, which happens to be in third position as a shoe exporter, does not even appear among the top ten countries when values are considered. So, both Asians and Europeans are good at exporting shoes but, while the former are unbeatable in the volume production and pricing (or perhaps we should say "producing below cost") of shoes that are simple to make and manufactured in large quantities the latter (Europeans) are good at making more elaborate products, with higher quality and with strong and recognizable brands, which are targeted at different market segments.

Having analyzed these figures with regard to exports, it is worth asking where all these exports go. So, how do import data compare with export data? World statistics for 2003 indicate that Eastern and Western Europe together account for a 38.5% world import share, while North America takes 31.5% of the global figure. The top ten importing countries are dominated by the United States, Japan and Germany as the three major importers of shoes. These are the countries which take the largest share of the yearly world production of shoes (in volume) and which absorb an enormous amount of footwear manufactured in regions such as Asia. A more detailed examination of the import-export flows would have indicated (not to our surprise) that the main outlet market for the billions of shoes produced in China is the United States. This is not unexpected if we considered that the tremendous shoe production growth in countries like China was due and in some respects planned as a consequence of the migration of shoe manufacturing to them from big countries like the United States.

More surprising, and certainly a reason for concern, is that if we look at the evolution of imports (growth rates over the past decades) we would notice a growth in the import of shoes in the European countries. For example, in traditional shoe making nations like Italy and Spain, shoe imports grew at between 8% and 10%, with the result that increasing amounts of Asian (mainly Chinese) shoes were consumed every year in countries with a long tradition in shoe making and a still-active footwear industry. In Italy, for example, which in recent years has managed to maintain its positive trade balance (difference between shoes exported and imported), for the first time in 2004 the amount of the shoes imported exceeded those exported, which created a negative trade balance. This occurred as a consequence of the progressive repositioning of Italian shoe production towards the upper segments of the market. However, the trade balance is still positive in value, and hence the picture that emerges is that of a footwear world made up of three major clusters, where each has a different view and all pursue different interests and goals. In the first of these clusters there are big producers and top exporters, with countries like China, India and Vietnam concentrating on maintaining leadership as major footwear producers and sustaining national economic growth. Despite some signs of a slow down in the rate of shoe production resulting in a more 'modest' annual growth of a few percent, this annual growth (in China, for example) is equal to the entire European production.

There are other factors to be considered: a slow but apparently constant increase in labour costs; the application of higher export duties; and the results of anti-dumping actions undertaken by European producers to force the Chinese to comply with international rules in terms of fairness of trade. But it will certainly take years before these factors have a tangible effect on the cost competitiveness of shoes manufactured in that part of the world. In conclusion, there is no evidence and no indication that Asia could loose its prime position in the near future as the "footwear factory of the world". The elimination of European import quotas at the beginning of 2005 gave another clear indication of how a hyper liberalized trade of shoes with no reciprocal actions to foster exports into China and the other Asian countries can worsen the situation. This can make the life of European producers more difficult, despite the growing interest and the unexploited potential of the wealthiest segments of the Asian countries for European shoes. The second of the three parts is mostly represented by the United States, with some European nations falling into the same group (Germany and the United Kingdom). The case of the USA is paradigmatic: in 1968, by eliminating almost completely every kind of import tax or duty on shoes manufactured abroad, the USA made a choice as far as shoe production was concerned: they decided to stop being a shoe producing country and restructured the sector to satisfy their huge internal demand (per capita shoe consumption in the country was 7.4 pairs in 2004, which amounted to 2.2 billion of pairs of shoes, of which 98.4% were imported, with 83.5% from China). In a few years their annual shoe production sank from more than 600 million pairs to less than 50 million and more than 250,000 workers had to look for different jobs. It was a clear industrial policy that the United States is still coherently pursuing and strenuously defending.

It is no surprise then that on all the occasions in which negotiations are undertaken (such as WTO, the Doha round and similar) to regulate international commerce they are against any decision to introduce taxes or duties that may limit the free and easy import of foreign shoes to the United States but also to other countries. The attitude of some big European buying groups, mostly interested in keeping high selling margins in a stable market by reducing to a minimum the purchase price of their articles, is not very different. It is not difficult to understand how this view of the market and this kind of attitude is a long way from the positions and interests of the third of these clusters.

This third group includes traditional shoe producing countries, which are striving to maintain a relevant share of their production and to avoid loosing their history and their knowledge on the product and its manufacturing processes. This group includes not only European nations such as Italy, Spain and Portugal but also Turkey and Brazil. All have been fighting to maintain their market position and to defend production, including creating trade barriers (import duties, quotas, antidumping fees and so on), as they face an invasion of far eastern products which menace their position in their markets. However, when they raise these kinds of barriers, it is not only against the Asian producers, they also find enemies among countries and organizations that are feeding big consumer markets (such as the United States, Germany and, in some respects, the United Kingdom) with shoes imported from the same areas. The interests of these countries are more likely to coincide with those of the Asian producers than with those of the other Europeans, which leaves the latter more or less alone to fight their battle. We need to ask, when the European shoe producers are actually fighting this battle in defence of their future, whether they are using the right weapons.

Therefore, on the verge of a new millennium, we can say that the European footwear business is dominated by the following critical driving factors:

- increasing competitive pressure from low labour cost producers
- excessive presence of fashion firms

Increasing Competitive Pressure from Low Labour Cost Producers

This is the main factor that has forced European shoe companies to split their production processes into several steps, to relocate each one of them in different countries, choosing production sites in those areas where low labour costs are

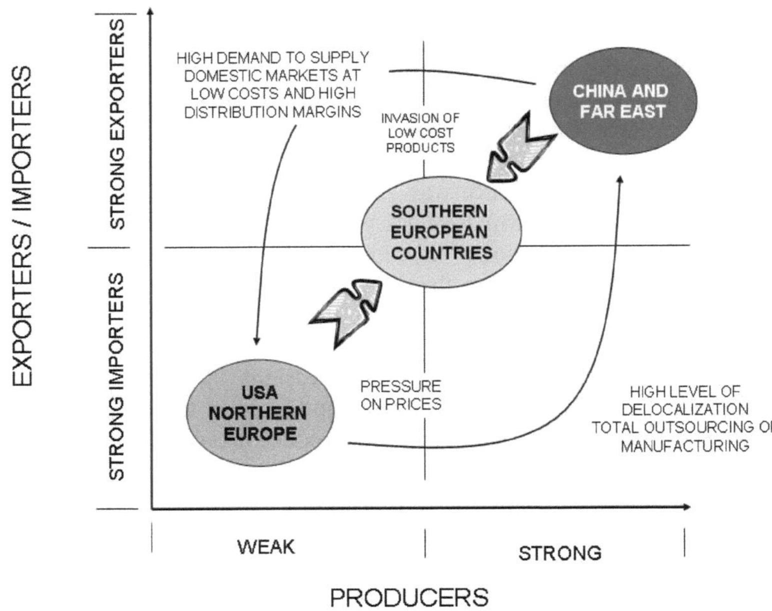

Figure 2.3. The three clusters of world shoe production

achievable and, as we have already mentioned, widely adopted as the easiest cost-reduction strategy to face the competition. This strategy has also forced the companies to outsource more and more steps of their production processes, always looking for the cheapest place to install new facilities; so the old local and regional clusters have been replaced by global, interregional and international networks. Companies have been forced to set up international production networks and complex organizations to handle them; a transformation that not all enterprises have been capable of mastering.

The second consequence of this increased competitive pressure was the push towards diversification, mostly in terms of progressive repositioning in the upper segments of the market and in developing a capability of supplying higher and higher quality products. When the high quality upgrading takes place, delocalization and outsourcing has to be limited, due to the need to preserve the quality standards (high quality means high local content of input; low quality tends to be followed by outsourcing of a large percentage of production abroad); this once more demonstrated how the delocalization strategy, which had to be adopted to counterbalance the increasing price pressure from the low labour cost countries, does not prove to be adequate when other competitive assets, such as quality, service, flexibility become relevant.

Excessive Presence of Fashion Firms

If the first driving factor mentioned above can be considered an exogenous factor (generated from outside the geographic area of interest of the European footwear companies), this second one is more of an endogenous nature. The increasing integration of footwear industry into the fashion industry has been dominated by a few multi-product oligopolies, which have exploited economies of scale and scope in activities such as distribution, marketing and branding across families of products. This has produced a concentration in distribution, a change in the production systems, a globalization of the production markets all supported by a dramatic spread of information technologies. As with delocalization, where producers tend to loose control of the manufacturing process, in this case they loose control of some other very crucial activities such as design, branding, marketing and distribution.

Caught between these two ponderous driving forces, European shoe producers (or manufacturers based in developed countries who want to maintain a relevant manufacturing basis in those countries) have to develop new strategies and to look for alternative market approaches to regain their competitive advantage. The ingredients of a recipe that could help Western producers to win back their market positions should include attention to the following: product quality; innovation in design and materials; flexibility of response to market demands; attention to individual consumer needs and provision of services rather than simply goods. Mass customization and product personalization involve most of the aspects mentioned above and can represent a repositioning strategy for this part of the footwear world. Its exploitation paths in footwear will be discussed in the next chapters of this book.

2.3 Mass Customization Made Simple (for Shoemakers)

Many different definitions of mass customization can be found in the technical literature and in the textbooks; it is not the purpose of this chapter to examine them in detail. A simple enough and adequate definition that is well suited to begin the analysis of its application to footwear is [4]: "Mass customization is the production of goods with a high degree of personalization with near industrial efficiencies"[7]. We intend to concentrate here on the specific ways such a "paradigm" is applied to the footwear business.

We used the term "business", instead of production or design, to emphasize the pervading effects that derive from the adoption of the concept of customization; as we will explain, all the processes that constitute the shoe product life cycle are heavily affected by such a concept and they need to be rethought, reshaped and sequenced in a different way. What customization is about basically is involving the customer, or better still the "consumer"[8] in the value chain of the shoe; keeping this in mind, in the following chapters of this book, every time we use the term "customer", we in fact mean the "consumer", using, in the context of shoe mass customization, these two terms as synonymous. There are different ways of

---

[7] p.3

[8] In shoemaking "customer" might have a misleading meaning, being the term mainly used by the shoe companies to indicate their clients, namely shops or retail chains, rather than the end users they build their shoes for

integrating the consumer in the value chain of the product, each one having a specific relation with the way the product is then developed, manufactured and sold (but not necessarily in this order).

In traditional mass production systems, end users (consumers) are not involved at all; the whole business is about manufacturing and *then selling* standardized products to customers who remain rigorously anonymous (the shoe factory does not know them individually; in general footwear companies tend to know very little about the real needs and demands of those who should be their reference consumers); goods are then *made to stock* (with all the related economical and cost efficiencies). Some "flavours" of customization can, to a minimal extent, be present also in mass produced shoes, but they all tend to satisfy classes or categories of consumers rather than individuals. We can have a soft customization when the consumer interaction point[9] is in sales or retail; this is the typical case of *match to order/locate to order* (which deals with the selection of existing standard products according to customer requirements) or *bundle to order* , for which existing products are bundled, based on situation of use, to customer requirements. Hard customization digs more deeply into the company structure and organization. In the case of *assemble to order* , that is assembling a customized product starting from standardized, pre-fabricated parts, consumer requirements affect the final assembling of the product, while for really *made to order* shoes, all components of the product are manufactured upon consumer's specifications and requirements.

We could also think (although the application of this model to the footwear field appears, at least today, more questionable) of having the consumer involved in the very first steps of the product life cycle (design and development), in what can be called *development to order* , in which he or she co–designs the product together with the producer, followed by a customized made to order. Then the term customization refers to changing parts of a shoe according to the needs and demands of a consumer. Breaking with mass and variant production, customized shoes are only produced when an order is placed by an end-consumer. Shoes are then assembled to order, based on pre-fabricated materials and components (the degree of pre-fabrication may vary) or completely made to order for an individual consumer.

We can ideally think of three possible vectors along which the customization of a shoe could proceed: **style/aesthetics, fit/comfort and function/performance**. Customizing the aesthetics of a shoe can be a relatively simple task, if we keep in mind that it does not mean transforming consumers into shoe designers (which they themselves don't want, as the outcomes of various market surveys indicated – see for instance [1]), but rather giving them the possibility of "building" or, more precisely, "configuring" their shoes by selecting their preferred option among a list of possible variants of the basic shoe design.

Fit and comfort involve more subtle and complex definitions (they are mostly defined by the last of a shoe, but also by the design of the upper, insole and outsole, the materials used in fabrication, *etc.*), which also imply a fair amount of subjective perception from the consumer standpoint; fit customization eventually deals with producing a shoe that consumers (and each individual consumer) will

---

[9] The point at which the consumers enter the value chain of the product.

comfortably wear without any need to accept compromises, should they not know the real size of their feet or the size required not be available in the shop.

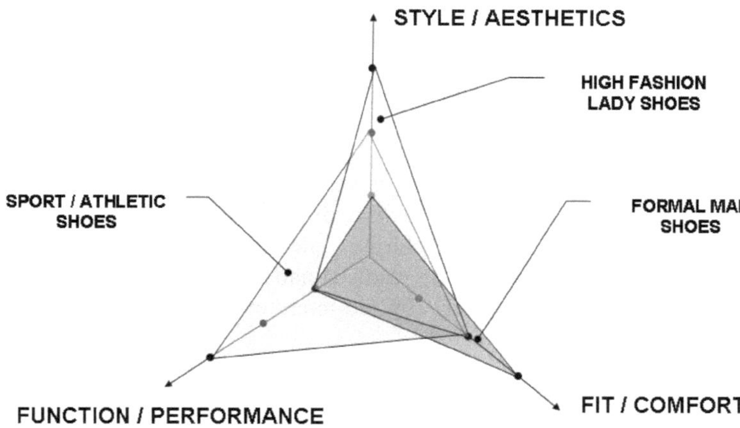

Figure 2.4. The three vectors of shoe mass customization

This way to customization does not only mean configuring the shoe from the aesthetic standpoint, but also applying some sort of dimensional adjustments to its building components. Then we have, ideally, functional customization, that is making a shoe for each individual consumer by "optimizing" its dimensional parameters, construction technique and materials in order to match the use consumers will make of their shoes (walking all day, driving most of the time, being seated at a desk and so on); this sort of customization, which implies a thorough knowledge of the biomechanical aspects of shoe - foot interactions, is for example used in sport shoes to enhance the performance of the athletes, but has not been yet thought of for normal consumers.

As the aspects mentioned above are those that can be customized in a shoe, such customization is then possible at different levels:

1. **Style Customization** - based on standard lasts[10] (and sizes) consumers can choose style options (colours, fabrics, leather, accessories) within constraints set by the manufacturer. This can be offered as a separate market option or be included in other customization levels.
2. **Best-Matched Fit** - the feet of each customer are examined (using devices called foot scanners) and matched to an existing library of lasts, insoles and soles with a much higher granularity than in today's mass production systems. Additionally, some style customization may be possible.

---

[10] A "last" is a sort of model of the foot, made in plastic, on which shoes are designed and built.

3. **Custom Fit** - the feet of each customer are examined and his or her specific habits are analyzed and used to make an individual last, insole and sole. Additionally, some style customization may be possible.[11]

Each specific level of customization has a well-defined impact on the three main processes that build up the shoe product life cycle (sale, design and manufacturing); we can measure this impact in terms of "gross costs" for the company to go for customization, where "gross costs" refers to investments in technology, manpower, organization, promotion of the new concept, retail disintermediation and so on. Table 2.4 shows this cost factor in relation to the three degrees of customization.

Style customization (first level) has a very low effect on manufacturing (it remains more or less the same as in traditional production, besides the aspect of having to handle more manufacturing variants), while design (more variants/combinations will have to be studied) and sales (a new sale model will have to be implemented, possibly by de-intermediating the traditional sale network) are more noticeably affected.

Table 2.4. Cost factors for three levels of mass customized shoes

	Design	**Production**	**Sales**
Style customization	High	Low	High
+ Best-matched fit	Very high	High	Very high
+ Custom fit	Very high	Very high	Very high

If we add best-matched fit to style customization, things get more complicated: the impact on design and manufacturing increases, because now the design department and the shop floor will not only be confronted with a higher number of model/style variants, but also, very likely, with a wider range of last sizes and last fits (which are needed to increase the chances of matching the size and fit of the individual consumers[12]). Also on the sale side complexity grows with the need to equip the sales outlets with foot scanners to measure the feet of the consumers coming to the shop, which also implies training the sales personnel, not only on the new sales logic but also on the use of "high tech" devices.

---

[11] Best matched fit is also called, in traditional terminology, "semibespoke" while custom fit is equivalent to bespoke.

[12] An average footwear manufacturer would normally offer its shoes in something like ten sizes (a measure of the length of the shoe) and just one fit (a measure of the width of the shoe); a good indication of what can be assumed as the minimum necessary offer for customized shoes, would include at least twenty sizes (full and half sizes) and at least three fits; this means a six-fold increase in the range of lasts and shoes to be handled in production.

When we move to custom fit in combination with style customization, we have a dramatic impact on all processes: the higher the level of customization the company aims at (fit/comfort, functional/performance) the more complex and high technology demanding the sales outlet becomes; on the design side new functions and operations become necessary, while on the manufacturing side it will be a real challenge and a quest for the utmost flexibility to handle a virtually "infinite" range (at least in terms of dimensional parameters) of products and to keep manufacturing costs of unitary lots at a level appropriate to the amount of money the consumer is ready to pay (more) for the customized shoes.

Eventually, all the variables of the complex equation that leads to the final cost of the product (the customized shoe sale price) must be combined in such a way that the result falls within that range of "mark up" to the normal selling price that consumers are ready to pay (various market surveys indicated that up to 20% more than the price of the same shoe made in the traditional way could be acceptable – see again [1]). It would seem evident and intuitive that producing in unit lots goes against the historical principle of economy of scale and will make the product costs increase; it is much less intuitive that a careful analysis of the entire value chain could highlight enormous and unexpected saving potentials that can compensate for increased costs in manufacturing. An analysis of this kind was conducted in the apparel field and presented at the First Congress on Mass Customization and Product Personalization [5]. Facts and figures presented in such a study, indicate that mass customization could actually bring down the final end user price (for example for a pair of trousers or jeans), by preserving the manufacturer's margin and even the retail margins. Similar results are to be expected for the footwear sector.

Having explained what mass customization is, how it can be applied to the footwear business and what it implies for the sale, design and manufacturing processes, it is worth understanding, in more precise terms, the way a hypothetical company offering customized shoes works. This is useful to highlight the kind of changes a traditional shoe company should undergo if it wants to adopt he paradigm of mass customization in its business operations. We can describe the operations of the company as one main process, broken down into five separate sub–processes:

1. Designing the customized shoes collection.
2. Selling the customized shoes.
3. Customizing the design for an individual consumer.
4. Manufacturing the customized shoes.
5. Delivering the customized shoes.

The whole process starts with the development of a new shoe collection "customer orientated"; depending on the company-specific parameters (shoe types – male or female, formal or casual, classical or trendy), the shoe models will be designed for the specific season and will take into consideration design and manufacturing requirements dictated by the design and manufacturing infrastructures of the company.

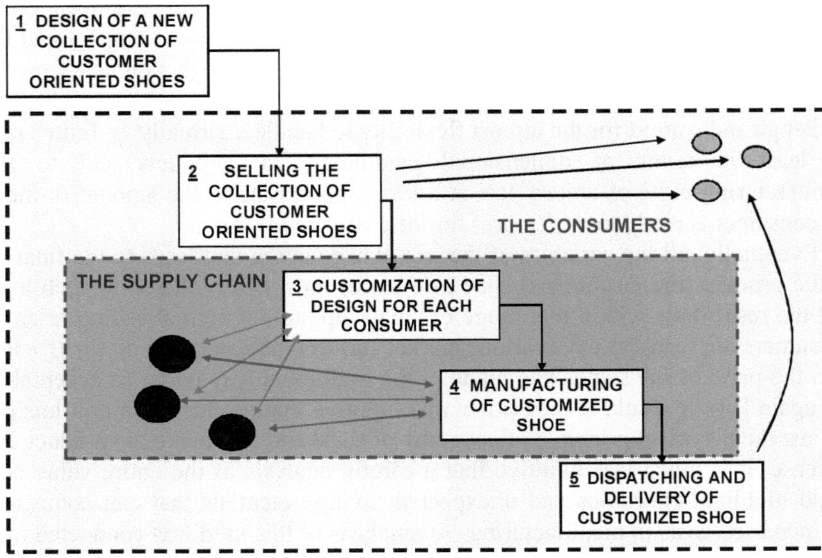

Figure 2.5. The five processes of the shoe mass customizing enterprise

To these typical contents of the design process (common to all footwear companies) a new dimension is introduced: customization. This implies additional requirements in terms of:

- *Material/component/colour variants* to be taken into consideration in the design phase and in the product structuring in terms of Bill of Materials; every "configuration" (design + choice of materials + choice of components) could generate a specific product code.
- *Higher granularity in terms of size and fit combinations* if the **best-matched fit** approach is chosen, which implies a careful coding of each individual last and of the specific size–fit instance. Nonetheless, size and fit definitions adhere to the traditional standards used in shoe making. The problem here is whether all necessary lasts should be supplied by the last maker to the manufacturing plant before production is started or if some of them can be ordered and procured only when needed (this could imply also producing the last internally in the shoe factory rather than at its supplier).

A further design dimension could be introduced here: biomechanics. As far as step 1 is concerned, biomechanics is to be intended as a means of improving the design of the shoe for the specific "segment" of population the footwear company intends to serve with its products.[13]

[13] For example if the consumer segment the company addresses to is composed of people who like a good cushioning of the sole and the company produces shoes with plastic soles, it

In relation to this aspect, the great potential mass customization presents for shoe companies stands precisely in the possibility, through the direct and continuous contact with hundreds of consumers, of learning very intimately what their features, desires and demands are and to make use of these findings to improve the design and technical contents of the shoes.

The next step involves the sales points. The "ideal" customer-orientated shoe enterprise bases its business model on three distinct sales outlets: brand or flagship stores, department store sections (a shop within a shop), and the Web. The brand and flagship stores are large and elegant, they exhibit the "image" of the company, they have an adequate staff and the most complete (and expensive) foot scanning equipment, they aim at the most exigent and demanding customers (for which the company will make truly custom made or custom fit shoes); the department store sections are shops within a shop where shoes are exhibited and sold , they are smaller with minimal staff and less expensive equipments and are aimed at capturing the largest possible share of consumers to the idea of customized shoes. The Web is the place for the future; it is already the privileged channel for the second purchase of the company's loyal customers. This is not a unique sales strategy in itself, but a winning component of a three-tier sales approach. Whatever the approach, the sales process comprises a precise set of actions and responses to a well-defined ritual that will be described later.

The various sale outlets will collect the customer orders daily and transmit them to the factory for their processing. If the company is structured to offer several levels of customization, orders are split into two groups: best-matched fit orders can be immediately processed and are sent to the shop floor, while custom fit orders will follow a separate path. In fact the "consumer specific design" has to be generated in order to start with the manufacturing of the various shoes. This step, that can be called "*design tailoring*" (or design customization for each consumer), and it requires the usage of the "quick adaptation functions" available in CAD systems and produces a unique last + shoe design that merges style requirements, depending on the model selected by the consumer, foot size requirements, depending on the morphology of the consumer's foot, and (possibly) biomechanical requirements related to the "function" of the shoe when used by the consumer. The output of this phase is a complete "project" of the shoe and of all its components, uniquely tailored on the specifications of the given consumer; CAD file data of the components are at the same time transferred to external suppliers in order to launch the manufacturing of all the various shoe components, although in the case of custom fit shoes the higher the level of process of integration in the factory itself, the easier is to serve the customers.

In the case of custom fit (or fully bespoke) shoes, only when the "customized" design is ready all the information required (in terms of quantities, production lots, machine part programs, material and component allocations) are available, and hence the manufacturing phase can start. The process at this stage will very much depend on the shoe typology and construction(s) adopted by the company; this dictates the configuration of the manufacturing line. Once more, in the most

---

will try to use "design rules" that correlate for instance sole geometry with cushioning effects, designing shoes with the appropriate cushioning properties

complex case of companies offering all levels of customization, two separate paths can be identified: *best-matched fit* shoes will go straight to the making and finishing lines, components (including lasts) are likely to be stocked in the company warehouses and production is managed in the usual way. Vice versa, *custom fit* shoes will flow through all shop floor departments (cutting, stitching, last making, shoe making and finishing) receiving those components that cannot be handled internally just in time from the suppliers.

As a last step of the entire process and to close the "consumer loop" (a procedure that starts and that ends with the consumer itself), the produced shoes are sent to the consumers, either to their personal address or to the sale points where they purchased them. This last step, although simpler than the others, implies some specific procedures and a good deal of integration with the information systems and infrastructures of forwarding agents and transporters.

Going one step deeper in the analysis of the operations of the shoe mass customizing enterprise, it is useful to understand more precisely what each step aims to achieve, which resources it necessitates and which are the constraints to be taken into consideration. The schematization we adopt here describes each one of the steps (or better sub-processes) of Fig. 2.5 as a "function box", whose task is transforming the given input into the desired output, using well-defined resources and under similarly defined controls, as follows:

1. **Design** - the design activity is performed, as previously noted, in a relatively traditional way: average information on the human foot, biomechanical requirements and style/fashion considerations all affect the design of the shoe and its last. These considerations are the *controls* that dictate the design work; the *input* is normally a "draft" (physical) last used as a basis for the work and formalized/non formalized style ideas, the *output* is the complete shoe project (including information for manufacturing), while human operators and CAD systems are the *means* to accomplish the work. This process is repeated for all the new models that form a new collection (although it has to be remembered that different style variants, or even different models, could share the same last). The main process steps are those typically undertaken by all footwear companies using CAD software to support their design activities. It is assumed here that the design cycle is mainly done "digitally" with extensive use of CAD/CAM systems and in particular of a 3D CAD; it is also important that modern and powerful CAD systems are available in order to increase the chances of having a "good fitting" shoe at the first iteration. It is also important that suppliers of "key" components (last, mould and sole makers) are equipped with computer systems that can communicate with those ones of the footwear company. To this extent, the approach of the mass customizing company is not that much different from the one of any modern and digitally equipped shoe company, but in this case the shoe mass customizing enterprise will make full use and take full advantage of its ICT infrastructures.
2. **Sale** - no matter what the sale channel is (brand shops, department stores or the Web), the sales process develops in a few basic steps: its *input* is

represented by the consumer in general terms (feet dimensions, habits and tastes), while the *output* is represented by information generated by the process (in terms of selected shoe design, feet geometric data and biomechanical patterns) and the purchase order. *Controls* are here represented by the available range of styles and style variants and the price brackets the consumer can buy in. The *means* to perform the activity are the sale assistants in the shops (or the consumer itself in the case of the Web sales), foot scanners, the product configurator (software) and consumer profiling questionnaires. Two major differences appear here, in the perspective of the shoe company and with respect to the way traditional ones are organized. The first one deals with the moment this process takes place: before production is started rather than after; normally shoes are produced and than sold to consumers. In the mass customizing case shoes are first sold and then made. The second major difference is that in this phase of selling, we look specifically at the consumer, at each individual consumers rather than to unidentified categories of generic buyers.

3. **Design tayloring** - this design phase is defined to be internal to the already mentioned consumer loop and it appears all the times that custom fit/fully bespoke shoes are offered.[14] It is in fact aimed at procuring the data to manufacture the customized pair of shoes the consumer has selected. The *input* to this activity are the geometric data of the consumer's feet, his or her biomechanical patterns, the product code corresponding to the particular configuration of the shoe selected and the administrative information related to the order. The *output* can be as simple as a "pure" manufacturing order (in the best-matched fit case) or as complex as the manufacturing order accompanied by a "customized" design of the shoe (the custom made last data, the adapted shoe design, the CAM data and so on). *Controls* in this activity could be the available lasts (in the best matched fit case), the required due date for delivery, the availability of materials and components, and so on, while *means* to perform the activity are again human operators (shoe customization technicians) and the appropriate CAD system.
4. **Manufacturing** - this relates to manufacturing the customized shoes, according to their specific sequence of operations. The *input* for this activity is represented by the "classical" set of materials and components (to be noticed that the last can be an input or an output depending on the specific scenario between best matched and custom fit), and the *output* is represented by the customized shoes. *Controls* are here mainly represented by the bill of materials/components, the machine part programs and the manufacturing routing for the various models. The *means* to perform the activity are production planning and management

---

[14] As a corollary when only best-matched fit/semi bespoke shoes are offered, this phase is not necessary, since no design tailoring for each individual consumer is needed; this is a very relevant affirmation since it indicates a possible path to customization that represent a workable compromise between consumers' satisfaction and complexity of operations.

software, machines, production systems and human operators. It should be noticed that it is very difficult to "generalize" this process since the more we go into details the more the description becomes specific to a given shoe construction.

5. **Delivering** - this takes place at the end of the entire sale, design and manufacturing cycle and closes the loop with the consumer; *input* for this phase are the produced shoes (each pair referred to a well defined consumer) while the *output* is here represented by the delivered shoes. *Controls* for this step are the consumer information (place of delivery, address, due dates and so on), while the *means* are represented by the forwarding and delivery infrastructures. It is a process typical of the customized shoe scenario, in particular when the final step of delivery to the consumer is taken care of directly by the shoe company.

The aim here was to decompose the operations of a hypothetical shoe mass customizing company in order to understand which are the relevant process steps, the order in which they appear, how they are related to one another, and which technologies and resources they imply. As a general comment, it can be concluded that most of the resources that this new business model requires don't seem to be that much different from the ones shoe companies already rely on. What is different is their relative importance, the moment they come into play, and the way they need to be managed. Some other technologies are more specifically connected to the business approach mass customizing brings in and for this reason they will be specifically treated and described in the following chapters.

Hence, a footwear company that is considering starting a venture in the field of mass customization should make a careful checklist of the resources it can count on and of their quality, in order to be sure that it is approaching the problem with an adequate level of understanding and consciousness. A possible checklist of the most important factors to be considered could be as follows:

- **Pervasive IT** - the key and the secret of running a shoe mass customizing company is a pervasive use of information technology (IT) at all levels and in all phases of the process. Mass customizing (MC) deals mostly with processing consumers' information in order to achieve consumer satisfaction; and when consumers are counted in thousands and for each one of them individual information (from specifications to individual orders) is to be gathered, processed and tracked, very efficient and powerful IT infrastructures are needed. Their role and relevance must not be underestimated, as its potential should not be neglected, because the knowledge of the consumer base that IT systems enable the company to achieve is one of the major values related to the adoption of MC. Tackling an MC project with no computer systems, manual process planning and control, fax and telephone is not impossible, but it will very soon limit the whole potential of the project and it will not allow a full exploitation of the paradigm.
- **Flexibility in manufacturing and in the supply chain** - relying on adequate manufacturing facilities with high levels of flexibility, capable

of handling the very unpredictable (in terms of quantities and typology) structure of the daily orders, is a must in order to cope with the two basic parameters that matter here: keeping the manufacturing costs as low as possible and duly respecting the delivery dates; and this of course with no compromises on the side of quality. Selling customized goods, and shoes more than others, is a business based on trust and confidence. If the trust relationship is not established or, worse, is broken, the venture is destined to failure. Where and how manufacturing is done, play here a relevant role, hence this choice deserves a great deal of attention.

- **Focus on the product -** we will further discuss this aspect in one of the following chapters, but it is worth mentioning it here. It should never be forgotten that at the end of the day, what is sold must be a pair of good shoes, or, more precisely, better shoes because the whole deal in shoe mass customization is exactly that one of offering a product with more contents (of material or immaterial nature) than in traditional shoe making. So if the product is not selected properly, accurately designed, well configured in terms of materials and colours, complying with the relevant fashion or style trends, it will simply not sell, no matter how well it is personalized.

As a conclusion of this introduction of the basic concepts of mass customization and of its background mechanisms, it is worth mentioning that there is no unique path to it, but that rather each company will have to identify and study its pattern to adopting the new paradigm. In general terms, what can be noted here is that there are at least two "models" that can be already identified and that indicate alternative ways of implementing the paradigm.

One model puts more emphasis on the customization side of MC and it finds its followers among new or traditional suppliers of bespoke shoes (typically for men) who want to modernize their handmade approach to the consumers, with all the most modern equipments and methodologies that computers and ICT (information technologies) are making available to them. Their products will than be an "affordable luxury", as the slogan of a very well-known producer of customized women's shoes states bringing down to a wider segment of consumers the "top class" prices that traditional purchasers of bespoke shoes are ready to pay. This is still not really for the masses though.

The second implementation of mass customization looks more at the masses of consumers, and is more adequate for traditional shoe producers, relying on their classical and available manufacturing facilities and really aiming at capturing a much wider range of consumers, who have never experienced customization, with a personalized product at a price that is only a little more expensive than the one they normally pay. We will further comment on these two models in the central section of the book where case studies of the current pioneers of shoe mass customization will be presented.

2.4 Three Good Reasons for Mass Customized Shoes

We have analysed in Section 2.2 how complex and challenging is running a footwear enterprise in these first years of the new millennium. Rules are changing, and the relative relevance of the various manufacturing regions of the world is largely different from what it used to be only ten years ago. Western companies are struggling to compete in the global markets, repositioning their products to more rewarding marketing segments, restructuring their production, trying to change their approach to business.

Mass customization and product personalization can provide an answer to these kinds of challenges, putting Western footwear companies in the position of moving the competition onto grounds that are much more favourable to them than the old ones in which production volumes and low costs seem to matter more than anything else.

In order to make a correct evaluation of the convenience of adopting the mass customization paradigm and to carefully assess the parameters that can determine the success of the project, several aspects must be thoroughly understood, both from the technical end economical standpoint. In Section 2.3 we have given some basic definitions and a description of the technical difficulties that producing and selling customized shoes imply, highlighting the implications of such a move and the impacts that it has on the organization of the company. These issues will be further exploited in other sections of the book, where a detailed presentation of the necessary enabling technologies will be given.

What it is necessary to address now is the very fundamental question: why should a footwear company choose to go for mass customization to better compete in the global markets in the years to come? Perhaps it is first worth clarifying which footwear companies we refer to in delivering the comments that follow. We have already stated that not all the footwear world is the same: there are regions that don't need (at least for now) special recipes to gain or maintain their commercial success, at least as long as there will be a market ready to absorb the enormous volumes of low cost shoes manufactured there; and we don't see in the short to mid term any signals that would indicate that this kind of demand for "mass produced" shoes should change that much. Then we are not thinking (yet) of countries like China, India or Vietnam (the manufacturing champions) when we propose "mass customization" as a path for a regained competitiveness.[15]

We are not also thinking of producing areas such as Latin America (and in particular Brazil), with strong domestic markets and a relatively protected position with respect to the Far East producers. The regions that we believe can benefit the most from the new business models and its market implications are: Europe first,

[15] Comments on a fast-learning and fast-growing country like China (and India too) are always very delicate; despite the common image of China as a typical mass producer, it must be noted that in the country there is a growing interest for all aspects related to technologies and in particular to mass customization and product personalization. So it has to be expected that it would not take long for the Chinese tiger to convert its production to the new paradigm, should the markets indicate a clear demand for that. So the time factor is once again very crucial.

with all its major footwear producing countries; North America, should it want to reintroduce at least part of the shoe production that it has lost with time; and, last but not least, Japan, a small nation, with highly demanding consumers and still a relatively well-established footwear industry. These are the countries (and regions) that we think could benefit by this paradigm shift and by the adoption of this more modern approach to the old-fashioned footwear business.

Let's return now to the fundamental question we introduced a few paragraphs before: why should a footwear company consider mass customization? There are at least three good reasons: the consumer, the product and the market. Enterprises producing consumer goods do exist exactly for these reasons: to serve consumers, located in a well defined geographical area and in a precise market segment with the most appropriate and adequate products. We will in the following pages explain how mass customization can find its motivation in relation to each one of these "drivers" and how it can help the company enhancing its performance in each one of them.

2.4.1 The Consumer

In traditional economic theories, consumers are "value destroyers"; they annihilate the "added value" that the company has put in the creation of the product by the act of purchasing and using it. When a traditional pair of shoes is sold to anonymous consumers, its life cycle, at least from the perspective of the manufacturing company is over, and there are no more value returns to the company during the "use" phase of the product life.

What mass customization returns to footwear companies is the direct contact with their final consumers; such a contact is now lost in favour of a mediated market approach in which the majority of the companies sell to retail networks which are then in control of the expectations, desires and feedbacks of the consumers. Needs and requirements of them are filtered back to footwear companies with time delays and through the "lenses" of the retailers. There is little use and few possibilities that such valuable information on what consumers want and how to better serve them can be made available to footwear producers when it would be needed. Even those companies which have direct sale channels (their own chains of shoe shops) don't exploit this possibility, since their sales approach is still, in the great majority of the cases, anonymous. Consumers are only known through their credit card numbers.

Mass customization puts the consumers back at the centre of the attention of the footwear producer, as it used to be in the past when the contact "consumer-shoe maker" was a direct, one-to-one contact; as it still is for the small community of manufacturers of bespoke shoes and their lucky and wealthy clients. Consumers are, in these kind of operations, greeted, inquired, scanned, measured, questioned, supported, and assisted in their purchase experience, transforming a simple necessity act (buying a good pushed by the need), into a ritual that makes the consumer feel that what the company is trying to do is not simply selling an item, but learning more about his or her desires and expectations. All these things greatly improve the purchase experience; they educate the consumers too, improving their capability to choose the right shoes for their feet and their tastes.

This is exactly how it should be: by "studying" the consumers the company "learns" something about them; they learn how their feet are alike, what are their fitting preferences, what their tastes and preferences are. There is a lot of value in this learning, as long as it is not dispersed but rather transformed into "knowledge" to be used to improve the design of the last and of the shoes in order to make them more and more comfortable and better value for the consumer, and to make material procurement forecasts more reliable and driven by consumer demands. In this respect mass customization can transform the traditional footwear making business into a "knowledge" business in which the knowledge of the consumers' biometry, preferences and expectations is the value that is created in the sale of a customized shoe. The consumer is no more a value destroyer but a value creator for the company that rewards him or her (yet at a cost, since customized shoes are more expensive than normal ones) by means of a more comfortable pair of shoes, perfectly fitting both in terms of measures, style, colours and materials that the consumer was able to choose based on his or her own desires.

It's a relationship of mutual trust that as long as the manufacturer is capable of maintaining its promises in terms of quality of the products, reliability in deliveries and attention in fulfilling expectations, it will create very loyal customers who will have no doubt and no hesitation in coming back many more times to maintain their relationship with the producer.

2.4.2 The Product

The product is certainly the second driver to go for when mass customization is considered. In the previous chapter we have stated that both designing and making customized goods, namely shoes, is more difficult and complicated than making mass produced ones. Although these complications can somehow be controlled and managed with the use of appropriate and well-suited technological tools (as we will describe in the following chapters), the fact remains that a "to be customized" product must be carefully studied, designed for modularity, and planned for production in order to minimize the impact that this completely different way of selling shoes brings with it.

Yet there are very relevant positive factors in relation to the product too that make mass customization attractive; first of all the fact that products are made only if needed. Mass customization of shoes implies that the company switches from a "made to stock" approach to a "made to order" one. Only those models, styles, colours and sizes that customers actually order are produced, as well as the tools and components that are needed for their production.

So the price that is paid in the product design and manufacturing planning phase, when a longer time and a greater attention is needed in order to approach these steps properly, is compensated by the diminished risk of investing money in tools, components, materials and shoes that will not encounter the favour of consumers and that will not sell as well as expected. It must be said that risks in these areas can be controlled and reduced but not eliminated completely. Initial procurements of materials and of the most critical components will have to be done in any case, based on forecasts and estimates of what the consumers will later want to buy.

A highly flexible supply chain (with suppliers capable of delivering components and materials almost in real time) and the availability of appropriate planning and procurement managements software tools can help reducing the risk and augmenting the advantages that can derive from a very reactive and lean supply chain.

Despite these difficulties (and the relative inefficacy of the tools that are currently available to support such operations), relevant advantages are still there: made to order means fewer or no stocks, virtually no low or under cost sales, and a lower risk of high quantities of unsold items that the company has to get rid of. It also means smaller sales surfaces; big inventories of several models in all possible sizes (so that the chances of serving all consumers are higher, no matter what their size or preferences are) are no longer necessary at the shop, with the direct (inventory cost) and indirect (cost for the area of the shop) costs that come with them. A relatively small sales surface where a few samples are physically available (whilst all the others are there only in the form of digital catalogues) is enough with all the positive implications in terms of reduced initial investments and of lower running costs.

2.4.3 The Market

Market motivations are the third area that mass customizing companies need to consider. Shoe companies are used to thinking of their market in terms of the "quality segment" in which their shoes would fit (high quality-high cost, competitive price-quality ratio, and so on), the use their shoes are aimed at (casual, formal, sport, fashion, safety, and so on), the gender of the consumers they are addressing (male or female) and the age category they belong to and, finally, the geographical area their consumers are based in.

This provides a very classical approach to the market, where the "market" the companies aims at serving derives from the intersection of the various specific aspects that we have mentioned above. Whatever its definition and composition might be, one point remains characterizing the common approach of shoe companies to the market: it is an indirect approach mediated by the retail network the company uses to bring its products to he consumers, as we have already mentioned.

The direct connection to consumers that mass customization implies turns into a transformation, for the mass customizing footwear company, from an indirect market approach to a direct one: the knowledge gathered during the sale phase, besides allowing the company to better assess the consumers' needs and to use such needs as "design drivers" for its new products (feedback on products), provides also an immediate and direct feedback on the responsiveness of each one of the "markets" tackled by the company. What actually happens is a strong identification between the "consumers" and the "market", being nothing more than the aggregation of many individual consumers, located in a given geographical area served by the company with its shops, who have common tastes for the kind of shoes offered by it and the same attraction for the personalization offer. In the traditional shoe business this identification between market and individual consumers does not, to this extent, exist.

For traditional shoe companies, markets are somehow abstract entities, with defined rules and mechanisms, not always well understood and mastered by the companies; markets, in the approach of the shoe company, have lost any relationship with their intrinsic nature of being eventually composed of individuals with individual tastes and interests.

Is this regaining of a more proper, human centric vision of the market, good or bad in itself? Can it be regarded as a competitive factor in favour of mass customization? On the one hand, even if we focus on markets as "collections of individuals", they will nevertheless posses their own dynamics and behavioural paths, the same they have when we consider them in the usual abstract way. Latin consumers behave and have tastes which are different from the ones of the northern part of Europe, Americans would be different from Asians, men would approach mass customization in a different way than women; differentiation factors are there anyway and they have to be carefully considered by the mass customizing shoe company.

On the other hand, in the case of mass customization and of a direct communication link of the footwear producer with its consumers, market dynamics, market responses and critical factors are obtained from a direct and daily observation of the sales record and of the individual choices of the consumers, or from punctual and detailed analysis of the reasons for lost sales or for unsatisfied consumers. What this regaining of the individual dimension of the market also brings in is a much shorter time span in determining the reasons for a missed sale or in highlighting the most relevant winning factors when a satisfied consumer walks out of the shop.

To summarize, we believe that mass customization enable shoe companies to regain a level of control on the market that is not easily achievable with a traditional organization, and this for the following reasons:

- **Amplification factor** - this is related to the fact that market responses and dynamics are now obtained as "summation" of the behaviours of a relatively high number of individual consumers, whose purchase decisions, preferences and wishes are collected in an analytical and detailed way; they are not "estimated" on the basis of forecasted average trends based on never consolidated (in a phase of rapid changes and very volatile purchase attitudes) historical data. This generates an "amplification" effect of those aspects that are more relevant to be considered to maximize sales and consumer satisfactions, whilst with the traditional sale approach those factors tend to be smoothed down rather than emphasized.
- **Time factor** – this is related to the almost "real time" feedback that the shoe company can obtain on its performance of footwear service provider (which is the right perspective to look at mass customization of footwear) thanks to the data communication link and ICT infrastructures the mass customization business rely on. Whatever works or does not work in the products, its quality, the way it is offered, its acceptance by consumers, is known and made available to the footwear company in real time, day by

day. This allows an unprecedented potential for quick response and adaptations.

These aspects together should motivate the potential interest of footwear companies for the paradigm of mass customization. So we think the following are good enough reasons to induce modern shoe companies to carefully look at the possibility of adopting mass customization in their business; to once more recall them:

1. **Consumer centricity** - the consumer is put back at the beginning and at the end of the life cycle of the product he or she buys, determining its characteristics at the beginning of the process and closing the loop when the shoes are delivered and used or becomes a value creator for the company, the value being the knowledge the company can gather about his or her tastes and needs. If the company can make good use of this knowledge by producing shoes that fulfil the expectations of the consumer, a long-lasting relationship of trust is built and a high level of customer fidelity is achieved.

2. **Product fertilization** - mass customized and also standard shoes can be greatly improved by the data base of biometric information obtained as a by-product of the sales operations. What is learnt from the consumers is translated into better specifications for the products and these improvements can be transferred also to the common mass produced shoes (should the company maintain both approaches) thus producing a generalized product enhancement, thanks to the fertilization that the knowledge gathered in the sales phase can produce to the entire model range of the company.

3. **Market reactivity** - as a result of the detailed knowledge of the mechanisms and the rules of a market that is known to be composed of a high number of individuals the company aims at serving, analytical knowledge together with real time information availability put the new shoe company in the position of being much more reactive to the changes in the demands of the consumers that constitute its market.

It is worth at this point analyzing the types of companies that can be potentially attracted by the new paradigm. We classify them with two parameters: their origin/history, which dictates their specific implementation path, and the kind of consumers they offer their products to. According to this classification, mass customizing companies can be described as *innovative shoe makers* or *appealed outsiders*, offering their products to masses or elites.

Innovative Shoe Makers

These are traditional/historical footwear companies that believe in the necessity of changing their approach to the market and are convinced that personalization and customization of products can represent an evolution of their traditional business and a way of gaining a definite competitive position. These kinds of companies can make use of available infrastructures (namely in product development and manufacturing) and often a network of retail points and sales outlets, in order to test new approach. They also have a pre-existing knowledge of the segment of consumers they want to serve and of the kind of shoes they might want to buy; knowledge that will be perfected and enhanced thanks to the direct approach to consumers that mass customization brings. Companies like these will very likely maintain (at least at the beginning) their traditional lines of mass produced products, taking full advantage of the fertilization effect that we mentioned earlier, running the mass customizing business as a distinct brand, a separate unit or even a newly formed venture (still utilizing the available infrastructures).

On the other hand, organizations like these are more difficult to transform and to become mass customization oriented; this requires a change in perspective of the management, some more or less in depth reorganizations of the business processes and, very likely, the updating of some or part of their technical facilities. At best, they have the possibility of exploiting the potential of mass customization, but will encounter many intrinsic difficulties in embracing the new philosophy easily and rapidly.

Appealed Outsiders

These are organizations or entrepreneurs coming from different sectors and different personal experiences who are interested in the paradigm of mass customization and who have, at different levels, an interest in footwear. These outsiders are favoured for the fact that they can shape and tune their organization and the related processes specifically for mass customization, without the constraints and the difficulties that their traditional competitors might find. If these are the positive factors they can count on, what they lack is the experience in the business (at least for some of them), a consolidated knowledge of the product and, more than anything else, they lack the specific design and manufacturing infrastructures they need to start up their ventures.

Typically outsiders of this kind will start from the sale side of the business and from a thorough concentration on the product aspects, which will define the failure and success of their adventures. The crucial decisions in this respect are those ones that will determine and identify the key suppliers they will need to begin with their operations, namely for the product development activities and for the manufacturing operations. In the central section of the book, in which several case studies of footwear mass customizers will be presented, detailed information about their origins, experiences and possible evolution will be given.

A second relevant decision for companies wishing to adopt the mass customization paradigm relates to the nature of consumers (hence the nature of the market) they want to tackle. One option is addressing their services and their products to the wealthy upper edge of the consumer pyramid. Offering their modern and efficient shoe customization services to an *elite* (yet in relative terms) of consumers; this kind of approach maximizes the quality of the offered product, the amount and the extent of customization, the efficiency of the delivery service at a price for the consumers that is not such to attract the masses. The sale/production volumes (in terms of pair produced per day) in this kind of approach will never be enormous and the size of the market served will remain relatively limited. The aim of this "variant" of mass customization is to modernize the typical artisan approach, by means of IT technologies, of some degree of industrial production, of "making digital" the key processes of the traditional producers of craft made bespoke shoes. Where is in this case the competitive advantage? It is in the possibility of making "luxury" affordable, if not to masses, at least to a wider segment of consumers hence with higher chances of obtaining interesting sale margins and rapid return on investments.[16] As we have already noted, there is much more customization here than mass.

When rather than elites, the company decides to serve *masses*, we have a true and full implementation of the mass customization paradigm; we believe that the greatest potential of the new business approach lies mainly here. The challenge that is taken in this case is the one of serving large segments of consumers, providing them with customized shoes at a price that must not be much more expensive than the one they are normally ready to pay for their regular shoes. It is in the adoption of this approach that adequate technologies, not only on the side of ICT, but also and mainly in the design and manufacturing processes, becomes necessary to allow a fast and efficient production of unit lots of very diversified shoes, taking advantage as much as possible of the economies of scale typical of traditional shoe production.

With the argumentations that have been presented in this chapter we have indicated some very general motivations that, in the opinion of the writers, should induce shoe companies or, in general, attentive entrepreneurs in moving towards the mass customization paradigm. What they have in common is the fact that they move away from the traditional considerations of competitiveness in terms of quality to price ratio and products content and of economical sustainability in terms of break even as a function of volumes produced.

What the new paradigm, more than anything else, introduces is a competition based on different parameters, it moves the battle to conquer the market onto a battlefield much more favourable for Western producers, in which the labour cost nightmare is much less a preoccupation than in the traditional business. The product alone is complemented by the service the producer is capable to offer to its consumers, sales margins are not strictly dictated by manufacturing costs and retail strategies; but they are more determined by the service content perceived by the consumer; the distribution of the added value in the various phases of the product pipeline changes too, allowing the company, also supported by the shift from a made to stock to a made to order production approach, to obtain higher margins on the terminal steps of the pipeline that compensate the higher costs that some others imply.

Besides that, two other major transformations actually take place when a footwear company decides to move to mass customization:

- *A transformation from a labour intensive to a capital intensive activity*, where capitals, material (machines, equipment and technologies) and

[16] We believe that this approach is more typical for outsiders than for traditional shoe companies.

immaterial (brand and product contents) assets count much more than labour. Capitals are much more easily found (and their cost is lower) in the western countries than cheap labour.

- *A transformation from a resource driven to a knowledge driven business approach*, in which knowledge at all levels (knowledge about consumers needs, tastes and desires, control on all the information exchanged in the various steps of the process, leveraging on such information to enhance the know how of the company) replaces the amount of resource the company can rely on, as the driving factor for competitiveness.

There is a last aspect to be considered in this analysis, which has been, up to now, mainly concentrated on footwear companies (or new start-up ventures) as such, and this is the role of retail. One might gather the opinion, from what was stated in the previous pages, that retail networks and retailers have no room in this scenario; indications were in fact given in the direction of assuming that the maximum advantage for footwear mass customizers would come when a direct control on the retail side of the business is there (so that the relevant margins that are generated in this phase can be fully absorbed by the company); this is certainly true, nevertheless it must be noted that retailers too can get benefits from this approach, both in the case of renewed partnerships with producers, and in the case of independent ventures they might think of starting themselves.

Retailers in this respect can take advantage of their direct contact with the consumers and of their in depth knowledge of the market, which is their daily field of action; it should not be difficult for them to figure out the potential for high quality service together with potentially higher margins, that mass customization brings. If suppliers are available with the adequate capabilities to produce customized shoes, an expansion of their product offer to include this typology of footwear would certainly give them a very valuable added value.

To conclude this analysis, we can state that mass customization, if properly implemented, can produce a winning scenario in which advantages are there for all the parties of the new business: for the consumers that take advantage of much higher levels of service (higher consumer satisfaction), for the retailers that can differentiate their product offer and aim at more interesting sale margins and, eventually, and for the producers, who have the possibility of more easily gaining a competitive edge on their traditional competitors.

2.5 Implementing Mass Customization in Footwear Enterprises

The aim of mass customized shoe manufacture is to produce individual units of shoes where each pair of shoes, including the left and right shoe, are different. To obtain the competitive advantages promised by the mass customization paradigm, both the organizational structure and the productive process must be updated [6]. The emerging tendency by a capital goods producer is to become a service provider rather than a product provider – according to the global service philosophy, which helps to make the whole scenario more homogeneous inside the mass customization context and its related requirements. Manufacturing companies, then, need to be ‘customer oriented’ rather than ‘product oriented’, consequently changing their own organizational structure. There is an identifiable trend to ask for manufacturing services (rather than production units) that could be rented only for the time of use; this is true for all SMEs, as they do not have great financial resources at their disposal.

Together with the development of systems, which are able to evolve during their own life cycle and adapt to market requirements, renting the desired configuration of such a manufacturing system may be a ‘financial’ instrument to mass customization. A number of different, new strategies are developing to deal with these market developments that require a market layout redesign. On the system producer side, the manufacturer of the machinery could retain ownership and lease ‘production hours’ or ‘products per month’, taking responsibility for operation, programming, service, maintenance, *etc.* Their customer (a material goods vendor), would pay for this service. As an alternative, a ‘system integrator’ might act as a ‘technology broker’, working as an interface and arbitrator between a company that needs a given productive capacity, and a group of functionalities and one producer of modular macro-components of the production facility [7].

This actor, probably supported by a finance or leasing company, would be responsible for the selection of the modules, their customization for the required process and their integration. They could rent a customized production capacity to the end user together with the operation and maintenance services. After completion of the rental period, the facility could be disassembled into its basic ‘building blocks’ and be reused again and re-assembled for a new production capacity that would fit new end-user requirements.

Expensive and time-consuming tasks such as maintenance and reliability become critical aspects in these evolutionary layouts. Maintenance would in fact be part of the service provided, so that the provider is interested in minimizing breakdowns. Thus the equipment must then be designed for maintenance. Modularity and re-configurability in manufacturing systems and system components must also be considered as key enablers for such a new market layout [8]. It must be noticed that in such a scheme the system provider becomes a process provider. This is coherent with the mass customization paradigm, in which, as previously described, the emphasis must be laid on the process and its life cycle rather than on the product. Many products are realized inside one process, and each process lasts longer then the product realized inside it. The type of update in the organizational structure of the enterprise for this market layout change can be considered as a part of the extended enterprise approach [7]. This seems to be a promising paradigm that allows enterprises, and in particular SMEs, to cope with the dynamic nature of the current global market and to compete with larger organisations [8].

2.6 The Role of Technology: Where to Find the Appropriate Enablers

Various technologies have been, in time, developed and tested to support shoe mass customization operations; they can all be regarded as “enabling technologies”, *i.e.* aimed at supporting the various processes (design, sale, manufacturing and distribution) through which the mass customization paradigm is realized in footwear. Most of them were born in the EUROShoE project (which will be extensively presented in the chapter 3), and at its conclusion most of them were at a relatively early prototypal stage. Nevertheless the work done in the project had the merit of highlighting they key technological components without which mass customization in footwear cannot work.

The aim of this section of the book is to provide the readers with a short review of the solutions that, in the years that followed the conclusion of the research effort, have appeared in the market and which shoe makers can rely on for the implementation of their mass customization projects. As the experience of the EUROShoE project taught, specifically developed technological elements are needed at all stages, from design to sales, from manufacturing to distribution; but what actually characterizes the mass customization approach is the way the shoes are offered and sold to the consumers (as it will very clearly appear from the case studies in Sections 4.2 and 4.3). Mass customization can exist even with “traditional” manufacturing means and with limited enhancements to the design procedures and the tools used in that phase; but without the set of dedicated equipments, software and procedures adopted at the point of sales, it simply cannot exists. Hence we will concentrate here on this family of technologies.

It is convenient at this stage to split the hardware and software “architecture” of a footwear mass customization system into two sections: the “front office” (or “front end”) part and the “back office” (or, better, “back end”) one. With reference to Fig. 2.6, we can identify four major “technological modules” in the architecture of the front end:

1. **Foot scanner** - this is device that is used to capture the features of the consumer’s foot and to obtain its more relevant measures. There are two classes of devices of this kind that can be used and that have been adopted by one or the other of the MC companies:
   - Manually/automatically operated measuring machines - this kind of system doesn’t actually perform a real scanning of the foot and does not produce a digital model of it, instead it registers the 3D location of selected points on the foot, from which distances, lengths, breadths and widths are calculated. The machines are basically manually operated and require a certain skill of the operator to deliver the desired results
   - Fully fledged foot scanners that, with the use of specific laser, optical or photogrammetric technologies, can generate a 3D digital model of the foot; on such a model several predefined geometric measures can be taken to characterize completely the biometry of the foot.

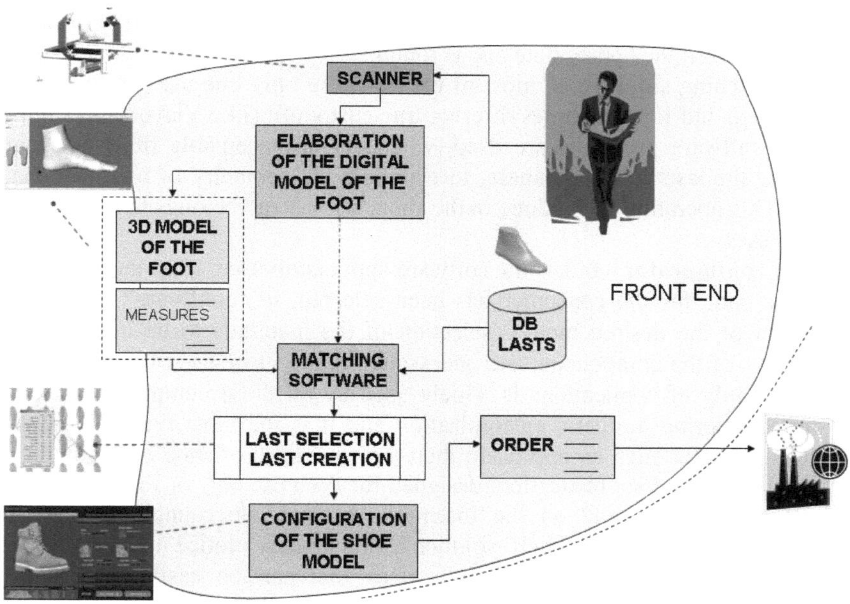

Figure 2.6. The front end architecture

2. **Matching software** - this is the second "core" module that enables the mass customization shoe sale process. It is normally a software module[17] that performs the comparison of the measures of the scanned (or measured foot) with the similar (homologous) measures of the last; the purpose of this operation is to identify the "best-matching" last among all those (for all models and all size/width combinations) stored in the data base of the company. Easy of use, reliability and consistency in the selections and automatic operation (no need of skilled personnel to use it) are the desired features of this software module.[18]

   There exist also other software applications that allow a visual and interactive comparison of foot-last. In such a case the matching software is not installed in the shop, but foot data are instead sent to the company's headquarters and

[17] Normally, but not necessarily, from the presentation of the case studies, it can be seen that in the case of Selve, for instance, the identification of the optimal last size for a given customer is actually done with the aid of no software; this step rather relies on the experience and know-how of the sale personnel.

[18] There are several options possible here: if the consumer knows in advance the shoe model and the style he or she wants to buy, the last "shape" is fixed by the model selected and only the size and the fit are to be searched in the matching phase (a "style-first" kind of search); when the consumer has no precise idea of the shoe model required, then a more general search is done that leads to the identification also of a possible range of shoe models (styles) that match his or her foot measurements (a "measurement-first" kind of search).

where each customer's order is manually processed, by visually (using another family of software applications) comparing the foot with the last in the database until the appropriate one is found.
The matching software is, most of the time, the only one used in the shop at this stage; but for companies offering true custom fit (like Viavor for example) other software programs are used, whenever an acceptable fit is not found among the lasts in the database, to elaborate the geometry of the tailor-made last. This operation is not done in the shop, but instead belongs to the back end processes.

3. **Shoe configurator** - this is the software application used, once the size and fit of the shoe for the consumer has been selected, to "configure" its personal variant of the desired model (selection of the materials to be used, of their colours, of the components and accessories that will be used for their shoes). This family of applications is widely used by all those companies that have started offering aesthetic customization and it is the most relevant enabling technology for such an approach; there are examples of shoe configurators at various levels of sophistication, designed for Web use only or for multiple use, both in the shop and on the Internet. Although we cannot yet speak of "standardized" off the shelf solutions, we already notice the presence of "dominant designs" for this applications that can be easily reviewed by visiting the Websites of the various companies that we present in Section 4.2. In this case too, it is not said that the configuration process can only be done with a software; once more the case of SELVE and also of the Mongolian BBQ show that the creation of the personalized variant of the shoe for the individual consumer can also be done physically with no software, by selecting the desired options from a component catalogue.
4. **Last database** - a database of lasts in digital format is necessary in the case the matching process has to be done automatically using a dedicated software module. This might not be an issue when the mass customization company is already using a 3D CAD software to design its lasts and shoes, but when this is not the case (as with shoe companies only using 2D CAD, or start-up companies which outsource the design process), this tend to become a critical point. In all these cases, lasts must be digitized, stored in a given and appropriate data format, and managed through a carefully designed database tool that can support easy and fast searches among numbers of occurrences that can be relatively large (hundreds or thousands of items depending on the number of different styles offered and on the range of sizes and widths each style is produced in).
5. **Order processing system** - although less relevant than the others, the importance of this last module should not be underestimated. It can take the form of a very simple procedure that, at the end of the scanning-matching-configuration process, summarizes all the relevant elements of the customer order and is then sent to the manufacturing unit (the back end). Or it can be a much more complex set of IT procedures, that integrate a POS (point of sale) system, that interface the company order processing and production management and planning system. Whichever the case is, this module completes the whole procedure, consolidates and stores the customer's data,

and initiates the process that will lead, eventually, to the delivery of the customized shoes to the consumers. Hence its role in the process chain must be adequately planned. It is a good and advisable practice that the procedure (or the software application) used can produce a set of printouts and documents that can be left with the customer to document its purchase and as a remainder of the choices he or she has made.

These five modules are found, with various degrees of sophistication and possibly aggregated in different paths, in all implementations of mass customization that we have monitored so far; they are then to be considered the basic ingredients for all projects of this kind. It is then important to examine what the market is offering, in terms of hardware and software solutions, to fulfill the needs of footwear companies wishing to implement their mass customization project, starting from the front end side of the business. In the next section of this chapter we will then provide a quick reference of the most consolidated available solutions that have appeared in the last few years. We have concentrated our attention in particular on *foot scanners*, on the *matching software* and on what we can call *integrated systems*.

Concerning the scanners, we will only provide information on automatic devices, leaving aside the kind of manual measurement systems that some companies are still using; although the reasons for their adoption are understandable (simplicity and low cost among others), we are convinced that they, on the other hand, require quite a high level of attention and skill in capturing the right points on the foot and only permit the acquisition of a few basic foot measurements. It is for these reasons that we believe that they are not really adequate to provide the matching application with the quality of data which is needed. Therefore we will turn our attention to the more promising and fairly mature automatic scanning systems.

Configurators are a family apart; as already mentioned there are no specific applications that can be bought as such on the market; all the big names presented in Chapter 4 have invested relevant budgets into the development of their own configurators (although the similitude of some of them leads to thinking of a common software engine underneath the different, customized user interfaces). And this is possibly the situation for all other companies which will have to consider investing, as far as this module is concerned, into the development of a tailor made software application to fit their needs.

There are different modes of mass customization as we have seen in the previous chapters and as it will be shown particularly in Chapters 3 and 4. However, we may say that when we talk in particular of footwear, we are confronted with the following technical challenges:

- the object to customize is extremely variable both in term of the foot as well as of the style designed and selected by the customer
- the foot itself may change slightly of shape during the day and during the seasons
- the type of shoe may vary greatly in terms of materials, components and form.

To customize the shoe it is necessary therefore to perform at least some measurements of the feet of the customer and then adapt these measurements to the type and style of the shoe chosen by the customer. The foot is a three-dimensional shape and therefore until few years ago only some more or less standard measurements were taken with some sort of simple ruler in term of length, width, girth, *etc.* If more of such measurements are taken, more information is gathered and therefore more precision on the real foot is possible. Of course the larger the number of measurements the more difficult it becomes to interpret and correlate the data. With the advent of cheaper computer power, it became feasible to scan the feet in three dimensions. Several technologies are available for 3D scanning: laser beam, 3D photogrammety, *etc.;* each requires a 3D reconstruction of the scanned data in terms of point clouds and/or plane patches, or 3D splines, NURBS, *etc.* Research in the past concentrated on the use of digital cameras as a scanner of 3D shapes.

There are several different techniques to make digital cameras. These techniques result in cameras with different quality and applicability to a variety of uses. The idea behind using cameras in a foot scanning system is to use low cost consumer digital cameras for the photogram metrical 3D measurement of the human foot. An understanding of the methods used in the camera's design is required to judge the suitability of a given digital camera technology for this specific application.

Fig. 2.7 shows the general schematic for transforming an optical image into a digital one. The image sensor is a grid of photo diodes which convert the photons that strike them into electrons. The electrons are stored in small buckets (capacitors) which are read out as a series of varying voltage amplitudes which are proportional to the image brightness at the particular picture elements (the pixels). The analogue voltage amplitudes are converted into binary numbers by an analogue-to-digital (A/D) converter and the numbers are stored and processed by a processor within the camera.

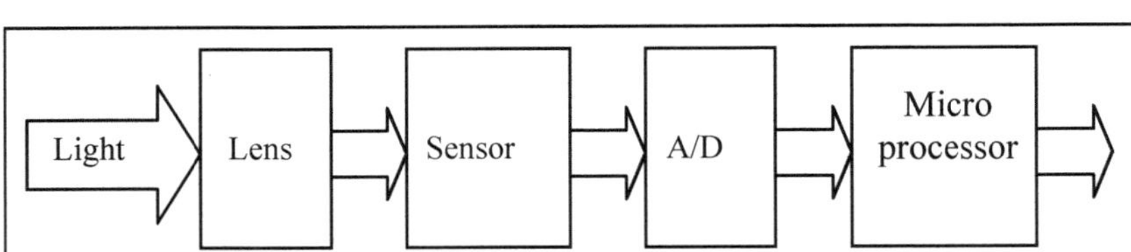

Figure 2.7. Transformation of an optical into a digital image

There are several definitions that are typical of digital camers:

- geometrical resolution
- dynamic range
- sensitivity
- blooming
- colour aliasing

We will not enter into the details because they are explained in several textbooks on the subject but the correct selection of the above parameters are essential for the definition of low-cost and easy-to-use foot scanners. The development of a very low-cost and mobile foot scanning system for capturing the relevant dimensions of the consumer's foot both in a shop and at home was one of the purposes of the EUROShoE project. The system was based on photogrammetry. The foot is covered with an elastic sock which is marked with special photogrammetric patterns. It is then photographed from a certain number of overlapping, but otherwise unknown, handheld camera positions. The photogrammetrical marks will be detected automatically using a combination of colour and black and white image processing. Once the corresponding marks in the different overlapping image pairs or triplets have been detected and the lists of homologue marks has been computed, the 3D coordinates of each mark are computed using close-range photogrammetry techniques.

The foot must be static during the imaging; more exactly, the foot can move in space as the camera does during the shooting of the different overlapping images, but it must not change its shape in between ("frozen foot" condition). This nevertheless means that the image acquisition time is limited to maybe 30 seconds.

On ground of the low-cost requirement the only candidates for the scanner were one-chip color cameras with mosaic filter. These cameras have adequate resolution but suffer from so-called "artifacts" (defects in the acquired pictures), especially blooming and colour aliasing. Although these cameras are not designed as geometrically accurate measuring cameras, they can be used for the foot digitizer having an accuracy requirement much below that of an industrial photogrammetry system. The more professional three-chip and high-resolution scanning cameras are much too expensive and not suitable for a consumer approach.

With, for example, a foot scanner like the light beam® 3D you are able to digitize the 3D shape of a foot and take - fully automatic - corresponding measurements like length, width and ball girth. Based on a patented MagicalSkin® technology, this scanner is also easy to transport and affordable.

The customer can access his or her 3D foot and measures immediately after a scan over a Webpage which can be integrated and customized into a Website. The client software controls the scanner connected to a PC via a USB 2.0 interface, it sends the data for 3D processing to the servers and receives the foot scan in about 20 seconds. The 3D copy of the foot can then be seen on the screen, measured and stored in standard 3D formats. Once processed, the data is available in a local database in the client software.

There are many scanners now available in the market and the most important ones are summarized below. Since technology in this area develops rapidly, we don't claim this to be a complete or up-to-date list of all the available systems of the this kind. It nevertheless gives a good idea of the range of solutions available at the time of preparing this book, their main specifications and possible field of application.

Scanners	
	**Corpus.e** The German-based company Corpus.e offers the Lightbeam© Photogrammetric scanner. It is a low weight, easy-to-use foot scanner that can be connected via USB to any desktop or portable PC. The consumer wears a special sock (the Magical Skin©) with a special pattern that is used by the software to generate a 3D model of the foot. No local processing of the scanned data is done; the frame captured by a digital video camera are sent to a remote server for the elaboration of the foot model. The company sells the scanner and charges an elaboration fee for each foot to be calculated. The software that controls the scanner is programmed to compute the most important measurements of the foot
	**Infoot** Infoot is a Japanese company that has developed the technology for the scanner that is commercially known with the same name; the system has been further integrated and developed by the UK company CSM3D (now part of the Torielli Group) which distributes it around the world. The scanning of the foot is done using a laser beam that is moved along its length. The foot to be scanned should (preferably) wear a white sock for the process. The reconstruction of the foot is done locally in the PC that is integrated in the system. The scanner has no built functionalities to calculate the measurements of the foot; additional software modules are needed for that.

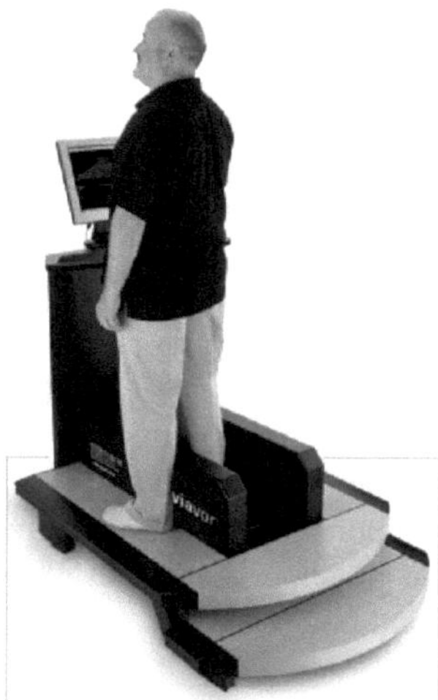	**Vorum** Vorum is a Canadian company with a long experience in the development of hardware and software solutions for the orthopaedi sector. The scanner the company offers for MC applications uses the same approach o the Infoot system (a laser beam that scans the foot); the elaboration of the foot model is also done locally in the PC that controls the scanner and that is part of the scanning station. The scanning software is very complete and can provide a full range of measurements on the foot.
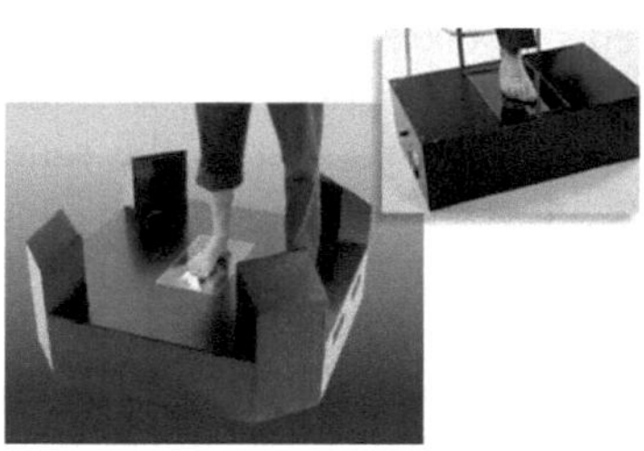	**FotoScan** The UK-based company Precision 3D offers a fixed base scanner (no moving elements) that uses a set of digital cameras to take images of the foot from various directions; these pictures are then combined to generate the 3D model of the foot using a photogrammetric reconstructing technique. The company also produces another device, a plantar scanner aimed at digitizing the plantar surface of the foot for the production of custom orthotics and customized foot beds. Although the equipment produced by Precision 3D systems seem very much orientated to the orthopaedic sector, they might also be successfully employed for mass customized shoes.

<table>
<tr><td></td><td>UCS
UCS is a young software and hardware company based in Slovenia, it is a spin-off of the mass customization experience of Alpina with whom UCS worked in close cooperation. The scanning system they offer is a very simple, entry-level equipment that can take, with a simplified scanning /measurement process a few relevant measures of the foot. Rather than a stand-alone solution, the UCS scanner is to be seen as a component of the integrated system USC has developed and that we will present in the next pages.</td></tr>
<tr><td></td><td>Formalogix
The company is based in the United States and has developed an integrated solution that has many commonalities with the UCS system. The scanner, which can also be offered as a separate unit, uses a set of fixed digital cameras to take pictures of the feet from various angles (notice that it is the only unit that scans both feet at the same time; all the others process one foot at a time). A 3D digital model of the foot is then generated; the basic length and width measurements can be extracted from the model of the foot, as well as many others of its relevant features. The model of the foot is stored locally in the kiosk and sent overnight to a remote server for the computation of the measurements.</td></tr>
</table>

A summary of the matching software available on the market is given next. If, for the scanners, we can already see a relative diversification in the product offering and a good level of "stability" in their technical solutions, for the last-foot matching software modules we are still in a much less developed situation.

Matching software	
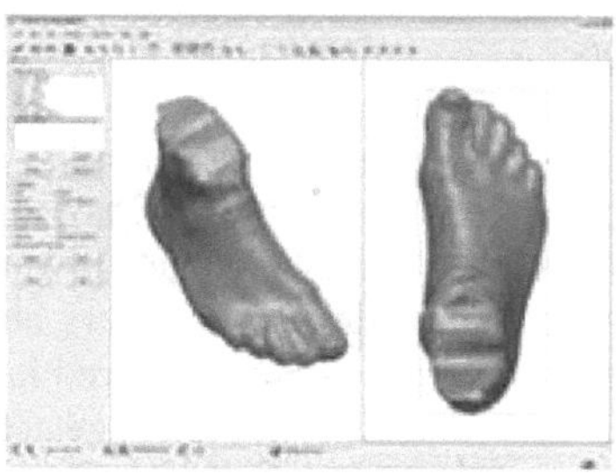	**FotoFit** FotoFit© is an interactive software application developed by the UK company CSM3D (developer of the well-known Shoemaster© shoe design CAD and now part of the Torielli group). The software is used for a manual matching of the foot against the last; the foot and the last are automatically aligned and then compared visually to verify the correct matching of the last. The software allows the user to perform a wide range of controls on the foot data and to make use of a set of predefined measurements.
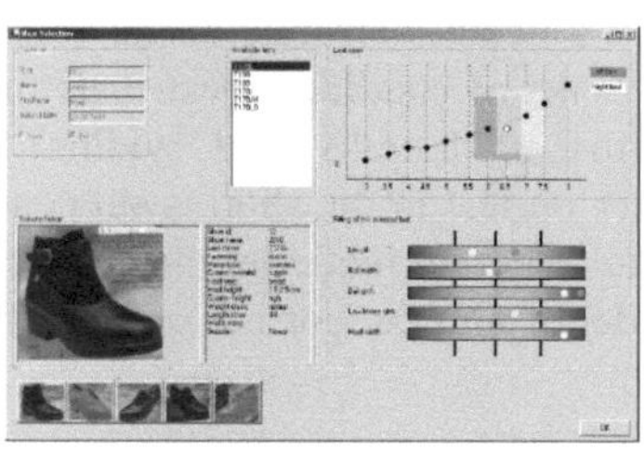	**Shoe Selector** Although this is not (at the time of writing these notes) a commercial product, but instead a prototype developed in a research project, it might soon become the first "off the shelf" matching software available in the market. It allows a "semi automatic" comparison of the foot data (measurements) with those ones of the lasts in the database and suggests, within a certain range of tolerance, the ideal (best-matching) last separately for the left and right shoe. A "control deck" gives the user a visual clue of the goodness of the matching. This application stems from the consolidated background in footwear technology of INESCOP, a well-known shoe research establishment based in Spain.

A list of integrated solutions is given next. These combine a scanner and matching software in one package.

Integrated solutions	
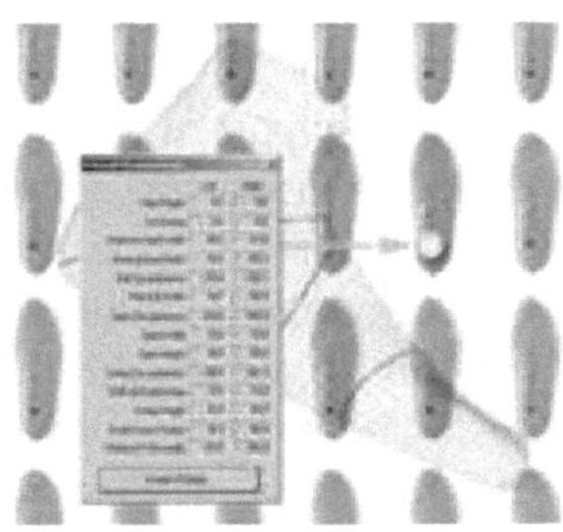	**Vorum** Vorum offers a complete and integrated solution to cover all the most relevant phases of the MC process and in particular a complete and powerful matching software that is sold together with the scanner. The software receives the foot data from the scanners and compares them with the previously stored last database. The best-matching last is suggested; when a best-matching last cannot be found, the system suggests to produce a custom fit. A different software module of the same suite of programs does the calculation of the tailor-made last.
	**UCS** The integrated solution proposed by UCS uses the foot data acquired by the scanner to perform an accurate search of the best-matching last; a ticket is printed at the end of the process. A list of candidate best-matching lasts (in terms of lengths and widths) is proposed; a try on test in the shop is supposed to confirm the selection made by the software and to help identifying the correct solution for the customer.
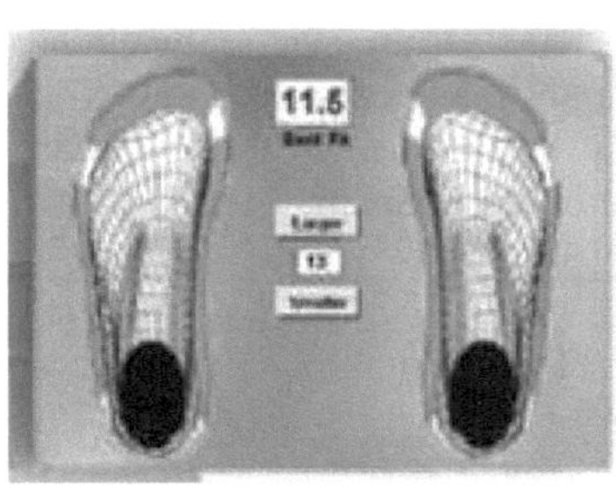	**Formalogix** The integrated solution proposed by Formalogix can hardly be classified in this group of solutions for shoe MC. In fact the company seems to promote its system more for shoe retailers (in general terms) than for shoe mass customizers. Its general validity as a system to scan feet and to select a best matching last from a library of digital ones to then produce a pair of shoes made on order, is certainly confirmed. But this does not seem to be a prime target in the view of its developers.

The particular application proposed by Formalogix (and also by UCS) can be regarded as a possible additional and promising "fallout" of mass customization technologies: instead of scanning feet and matching lasts to produce shoes on order, the idea is to use the same approach to help consumers and retailers in choosing the right size of shoe.

These companies offer their solutions as "virtual fitting systems": the client is supposed to indicate the specific model of shoe he or she wants to buy, then goes to the scanner to get his or her feet scanned and the system will specify the correct shoe size. Perhaps we are not dealing here with mass customization (shoes are already there, they are not done on order, there is no involvement of production), but it cannot be denied that this approach offers clear advantages to consumers - it educates them to choose shoes in a more "professional" way and it gives the retailer the chance of offering its customers a valued added service. No mass customization but certainly consumer centricity as a fertilization from the pure mass customization approach.

2.7 Mass Customization, Footwear and Economics: A"Win all" Game

The implementation of mass customization (MC) principles in a mass sector like footwear requires that shoe producers change their vision on production and organization. This is already taking place at the level of large companies, which are offering the possibility to their customers to personalize the product (Nike and Adidas are applying the principles of MC). In the case of small and medium size producers this is much more difficult due to problems in adapting processes and production, which are designed to produce large amount of items.

The organizational re-engineering is one of the most important issues for the development of a mass customizing company, because different actors (shoe manufacturer, external designers, suppliers, component manufacturers, subcontractors and customers) are all involved in the shoe life cycle and their relationships need to be reengineered. So MC is not only a matter of new production lines or new machines but it is also a matter of redefining [9]:

- the internal processes of the footwear company
- the processes with external partners, suppliers, subcontractors, *etc.*
- the processes with market/customers (sales and distribution)
- the implementation of ICT tools to appropriately support processes.

Companies that are adaptive to the changing environment to specifically meet the customer demands/expectations are more likely to be affected by the benefits of MC. Taking as a model the Dell strategy, a company should eliminate some intermediaries in its production chain and interact directly with its customers especially via the Internet [10] to provide tailored solutions supporting a physical product with value added services. It may be a high tech or a manufacturing product but it has to be customized. Improvements in the relationship between production and marketing within the firm is another way to get benefits, while internal and external flexibility of processes should be enhanced also by the introduction of new machines and new ICT solutions.

The aim of this chapter is to study different strategies for traditional shoe producers that are willing to implement different degrees of mass customization. The work is based on a strategic and economic evaluation of the different possibilities offered to a shoe producer willing to apply such an innovative paradigm. The evaluation is based on the data collected from the outcomes of research carried on in this field and on data from real traditional companies producing shoes. Simulations have been done to define costs and new working time for new strategies implementation and they have been tested to verify which were the differential costs that a traditional company has to sustain in order to produce MC shoes using innovative machines and using ICT platforms to integrate the production process with all the other processes of the shoe life cycle such as sales, design and distribution [11].

Until now in literature many theoretical references explained changes in the cost structure for implementing an MC strategy [1] but few economical estimations are based on real data. The data can be considered a preliminary analysis of the most important costs and benefits of such an implementation.

2.7.1 Most Important Changes in Terms of Costs and Benefits

As has been emphasized in the previous sections, the MC approach implies organizational and technological changes at company and supply chain level. Summarizing some points already described above, it is possible to asses how they can affect the typical production cost structure [12, 13, 14]:

- The costs increase for lower economies of scale as product variety increases and lot size is reduced;
- MC influences order management, moving from ‘make to stock’ (MTS) to ‘assembly to order’ (ATO). This guarantees a decrease in stocking costs since warehouses are no longer packed with shoes waiting to be sold. This implies an increase in the cost of quickly coordinating shoe assembly.
- There is an increase in set up costs due to the diversification of products (many changes of tool, *etc.*); in the case a company decides to produce MC shoes with a new dedicated production line this category of costs is counterbalanced by the higher flexibility of the new production system.
- An increase in diversification costs can be counterbalanced by the postponement of the phases more affected by customization along the production process.
- There will be changes in the distribution costs, because the direct effect of an ATO policy is an increase in lot number which means an increase in transportation costs due to more frequent and more spread delivery.
- There will be relevant investments in automating the production and assembling process by buying innovative and adaptive systems.
- New competences will be required to make use of new automated machines in addition to the consolidated skills at production level, which means investments in training and upgrading.

- Finally investments in software for production management and customer relationship management will be required.

In general, additional costs need to be counterbalanced by appropriate capability of design for variety, developed thanks to training and hiring new designers, use of a modular product architecture, increase in the capability of production planning and control with the support of appropriate tools and postponement strategies [1]. Most important is to increase the capability to integrate customers during the value creation process starting from re-engineering of the sales process and the relationship of this process with production, as already emphasized.

Another important factor to consider is the cost of materials (raw materials and components) and in particular the cost of the shoe last (the last is the block of wood or plastic shaped like a foot, used for the shoe assembly). For standard shoes the request of new lasts is mainly linked to the seasonal new fashion and they can be reused many times as long as models are the same and the last is not ruined by multiple steps of processing on the lasting machines. In the case of customized shoes the last can be personal for each customer[19] and even if it can be possible to reuse it for more than one pair of shoes (if the shape is the same), it is necessary to consider the higher amount of sizes and fits which are to be available to satisfy all customer needs. The last is already a relevant cost for the shoe producer and the production of customized shoes is linked to the customization of the last which increases a lot the costs for materials both in the case the company decides to make it internally or to outsource its manufacturing. In the former case it is necessary to buy innovative machines for last production which have still a limited capacity and can be the bottleneck of the whole production, while in the former case it is necessary to outsource a higher amount of lasts.

As we have listed above, the approach to MC has also a significant influence on lot sizes, reducing the average size of shoe lots from 500 - 1000 pairs to about 10 - 20 pairs per lot (where each pair of the lot can be different from the other), or even to strictly one pair lots. This can have an impact on many other costs like production and distribution costs, increasing the complexity of planning, scheduling and tracking orders [15]. These expenditures can be counterbalanced introducing the usage of efficient ICT tools to provide the required autonomy in decision making and flexibility in job sequencing and to increase efficiency of the distribution network.

The costs of stock management (raw material, [WIP] and final product) for customized shoes decreases compared to standard shoes only if a lean management of materials is applied to eliminate unsold finished product. In fact it is possible to consider that the amount of stock of raw material and WIP (uppers and components) decreases when a strong investment is undertaken also on the ICT systems supporting the capability of the company to manage production and sales in an integrated way, reducing the amount of stocks at the end of the period compared to the amount stored for standard shoes.

---

[19] This is true for custom fit or bespoke shoes; it is not the case for the bestmatched ones.

2.7.2 Steps to Mass Customization Implementation

Given these changes on the cost and benefit structure, it is now possible to evaluate the impact of MC on a traditional shoe factory. Typically shoe producers completely outsource the preliminary phases of production, like cutting and stitching and the manufacturing of main components as heels and soles.

The investments in MC can be implemented at different levels according to the degree of technological and organizational changes the company wants to apply to its own production system and organization. It is assumed to deal with a traditional scenario where a shoe producer purchases from subcontractors raw materials and components to obtain the final products. The most important investments which should be undertaken by a shoe producer in order to implement a MC strategy, have been grouped into five *macro cost categories* (progressive tiers of cost allocation on the way to the full achievement of the MC project) according also to previous studies [16]:

Macro 1: Purchasing a consulting service for MC from a specialized centre
When a traditional shoe producer decides to start with MC, it may need support of external consultants on how to transform its business to MC. This service supports the company in defining, for each phase of the new shoe production, most convenient times and methods for reengineering processes through feasibility studies. The service may consist for example in the realization of a simulation analysis based on the current production configuration of the shoe producer itself considering both standard and customized shoes. In fact the company does not have to necessarily change its manufacturing processes but it can simply modify the production organization and the industrialization of the customised shoes. The quantification of the cost of the service provision are related to the time required to the technical staff of the service provider to collect all the data about the production system, to implement the simulation model, to run the simulation and to brief the footwear company on the results of its study. This first source of costs related to the implementation of a MC project, can be regarded as the initial step – minimum investment for the exploratory phase of the project itself.

Macro 2: Changes in the sale process
This investment category represents the first concrete step towards MC; it supports the direct relationship between the shoe producer and the customer. It implies the purchase of a foot scanning system which will be installed in shops to directly gather information on customers' foot size and style preferences. The company can decide to purchase the foot scanning systems for all or some shops selling its shoes (they can be both private and multiple label shops according to the strength of the shoe producer in the final market). The investment includes also the software (databases and data gathering) to acquire and store data from the shops on the customer's requirements, the training of personnel and marketing activities.

Macro 3: Software purchasing
MC requires short response time for very small production lots (there may be also lots with only one pair of shoes). This means a very high degree of integration in

the supply chain. The "shoe pipeline" includes many different actors which should integrate their processes using software specifically customized for the shoe sector: an integrated environment where CAD, CAM, ERP, SCM and a scheduler are necessary at this stage to communicate and transfer information also to subcontractors to shorten their response time.[20]This will imply also investments and costs on training personnel. Moreover it implies assuring a proper time to market for products. This can be obtained mainly managing production in order to avoid delays and maintaining production time similar to standard shoes and organizing distribution in order to deliver the product to the customer.

Macro 4: Purchasing of new production machinery
The company can, at this stage, decide to introduce innovation at production process level by adding new machines for the customized shoe manufacturing. This change can involve the whole process (from cutting to assembly) or just some steps of it. An example is represented by last making which is particularly important in the MC approach. The last of the shoe needs, in the case of custom fit, to be personalized for each customer and outsourcing its production can be very expensive. Usually during one season a shoe producer can reuse the lasts for 300 - 400 times but for a MC production it is necessary to increase the number of lasts according to the level of personalization adopted.

Macro 5: Purchasing of a new internal logistic system
The company can, eventually, decide to invest also in a logistic system which can be devoted not only to the storage of shoe components , both processed internally at the plant and externally by suppliers, but also to dispatch such components to different locations both in the stitching and in the making departments according to the planned schedule [16]. This investment can increase the flexibility of the overall production system and the capability to handle urgent orders.

2.7.3 How to Integrate the Macros Towards Mass Customization

As it was explained in the previous sections, there can be different kinds of MC depending on the degree of customization: "Best-matched fit" (BF) are shoes with a wider range of fit and size compared to standard (STD) shoes, so to meet the needs of many different customers matching between available lasts and foot measurements; "custom fit" (CM) shoes are more advanced because the last of the shoe is personalized for each customer: CM shoes are also considered for particular categories of customers with problems on foot shape. Also in the case of CM shoes, once the foot is measured by a scanner the software first tries to create a match between the available lasts for BF shoes and the customer's foot. Usually as long as the foot is not peculiarly shaped, the demand can be satisfied with the available lasts without the need to produce a specific one, which would add additional costs to the whole process.

---

[20] A complete suite of programs of this kind has been developed in the EUROShoE project; for further reference see [5].

Responding to the market demand of customized shoes (either BF or CM or both) implies combining the five macro investments described above in relation to the different degrees of MC to be implemented. According to the demand level and to the company investment capabilities, this work investigates the impact of each investment scenario on a medium size, traditional shoe producer to understand how a shoe producer already in the market can combine traditional and mass customized production using outstanding enabling technologies under company's constraints on production, organization and resources.

In the case of macros 3 and 4 it was necessary to consider different sublevels of investment. They are in fact the most challenging ones because they include the implementation of the most relevant new technologies. This means that the company can step towards MC without introducing the whole changes required by the macro but just some of the steps. In particular:

- Macro 3 is structured into three levels:

Level 1 - introduction of a database for customers' data management to use all the data gathered from the shops during the sale process for the design and production process.

Level 2 - introduction of CAD and CAM systems for the last and the shoe design and production phase.

Level 3 - introduction of ERP, MES, PDM, and SCM[21] systems for integrating the different company departments and the external partners.

Table 2.5. Macro combinations

		Macro 1: service	**Macro 2: sales process**	**Macro 3: ICT**	**Macro 4: production**	**Macro 5: logistics**
Higher degree of innovation from strategy A to strategy D	**Strategy A**	Feasibility study from service provider	Scanning tools	Level 1: database for managing customers' data		
	Strategy B	Feasibility study from service provider	Scanning tools	Level 2: CAD and CAM systems	Level 1: machinery for lasts production	
	Strategy C	Feasibility study from service provider	Scanning tools	Level 3: ERP, MES, PDM, SCM systems.	Level 2; new production line	
	Strategy D	Feasibility study from service provider	Scanning tools	All software	All new production machinery	New internal logistic system

[21] ERP stands for enterprise resource planning; MES is manufacturing execution system. PDM is a product data management database and SCM a supply chain management system

- Macro 4 is structured into two levels:

Level 1 - introduction of machinery for last production to produce internally mass customized lasts and shoes.
Level 2 - introduction of a new flexible production line fully automated to change completely the production process.

Table 2.5 represents the way the different macros are combined in four strategies towards innovation to respond to MC requirements. These strategies are somehow incremental; starting from the minimal investment required for MC to the most complete level of investment. For each strategy additional costs and benefits for a traditional company are evaluated and estimated as shows Table 2.5.

Strategy A
This scenario represents the very first step that a company can take towards MC. It can be considered the minimum investment in order to produce best fit (BF) shoes. In order to apply this strategy, the company outsources a feasibility study to a specialized service provider not only for the process but also for the product re-design. The other relevant aspect is to assure MC application is the re-design of the sales process: each shop, licensed to sell the brand, must be equipped with the scanning tool (a foot scanner of some kind) and the software for the customer's information management. The investments for this scenario are the purchasing of the scanning tool which includes the matching software and the configurator (macro 2). Moreover in this context it is necessary to introduce a Database for the management of customer information (which corresponds to a first level of innovation related to the macro 3). Costs are related to the feasibility study outsourced and to the equipment for the shop (including maintenance, training, and so on). Marketing costs are also relevant and very important for this and all the other scenarios and they have been taken into consideration in the analysis.

Strategy B
In the case that a company wants to start the production of custom fit (CM) shoes, it needs to add to the previous scenario the cost of machinery for last production. For standard and BF shoes, lasts production is outsourced as in the traditional shoe process. The fact that the company produces lasts internally implies that it purchases also the CAD and CAM systems (macro 3) to integrate the shoe design process and the part program definition with the production of the lasts. Other relevant costs to consider in this scenario are for training of employees on the new software.

Strategy C
In this case the company is more innovative, so it decides to further improve both the production process and the IT infrastructure. The company installs the new automated production line, dedicated to the customized shoes (BF and CM) and it hires new workers for the cutting, stitching (which becomes an internal activity) and for the assembling process. The assumption is that the company sets up a new manufacturing line and that the production of standard and customized shoes is run in parallel on different lines. The IT level is improved adding the ERP, MES, PDM

and SCM systems in order to manage and optimize the information flow concerning products, customers and suppliers. Other costs that will be necessary to implement this scenario are related to licences for software updating, maintenance costs for new machinery and labour cost.

Strategy D
In this scenario the company makes the whole investment for MC adding to the new automated production line, an innovative logistic system to manage the distribution of materials and components along the different machines and along the phases of the new production line. The costs related to this decision are related to the acquisition of the logistic system, pallets, maintenance and operators' training.

Implementing any one of these strategies means that it is necessary to consider also the costs for external logistics for delivering customized shoes. This cost is higher compared to the standard production, because the company uses carrier services to ship everyday shoes to shops or to final customers.

2.7.4 Hypothesis on Shoe Demand and Selling Price

Once the scenarios are defined some hypotheses for the test cases and some basic characteristics and values of the variables used in the evaluation need to be fixed. Demand for customized shoes is very difficult to assess. Some preliminary market analyses have been conducted at European level forecasting data and trends concerning both male and female expectations on customized shoes [1].

A first hypothesis done in this work is that the demand for customized shoes equals the production capacity of a shoe producer and no costs of unsatisfied demand are considered. In this study we do not take into consideration problems related to capacity limits in satisfying demand for customized shoes. It is assumed that the traditional shoe producer has one traditional production line with a capacity that is not fully used for the yearly average production. This is the most of the time true since in this sector there are seasonal peak demands to be satisfied and the shoe producer always needs to have some spare capacity. This capacity can be used to produce customized shoes in the case the strategy is very conservative and the management decides not to buy the new production line for MC. In the case of customized production this capacity can be used for BF shoes; the peak of demand for standard shoes can be outsourced if necessary.

As far as figures are concerned, the hypothesis made in this work is to consider a shoe producer with a plant with one production line that can handle a demand of around 100 pairs of customized shoes per day. This is also linked to bottleneck problems given by some machines and some phases of the process. In fact, in considering the first step of a customized shoe production, represented by the milling of the lasts, it is necessary to keep in mind the throughput limit of such machines which still represents a constraint for the process. If the company decides to outsource the production of lasts for CM shoes, the cost will increase proportionally with no clear advantage with respect to the purchasing of the machine and for this reason it is advisable to the management to keep the last making process internal.

Usually standard shoes, even if sold in big numbers, do not completely satisfy the whole population of customers but just the mean. In fact, supposing that the tastes of the customers have a normal distribution, MC permits an increase in the number of people fully satisfied by the product and an increase in the consumer loyalty ensuring a long - term fidelity to the brand. Mass manufacturing fails in responding to individual needs regarding the desired ideal product of individual customers [17].

From the graph in Fig. 2.8 we assume that the different degrees of innovation also imply different capabilities to answering market demands. If the level of innovation is low (scenario A and B) the producer manufactures customized shoes (BF) using the traditional production line. Over the years the number of standard shoes manufactured could be reduced in order to produce more and more customized shoes; however in scenario A the company doesn't only produce CM shoes; so it is assumed that in the first years the company can take advantage of its spare production capacity, but that this capacity is very rapidly saturated so that, from year 2, there is no increase in the quantities of MC shoes produced.

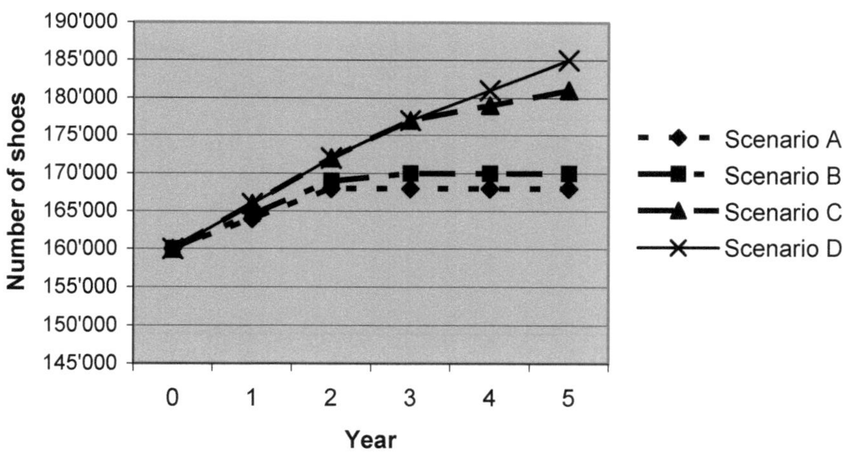

Figure 2.8. Total demand (production volumes in pairs per year) for each scenario

In scenario B (Fig. 2.9) it is assumed that the volumes of CM and BF shoes constantly increase with time, initially relying on the spare production capacity; as the market for MC shoes takes off, the production of standard shoes is progressively decreased to accommodate more and more customized footwear. If the level of innovation is higher (scenario C, D, see Fig. 2.10) there is no reason to reduce the production of standard shoes, because the customized (BF and CM) ones are produced on the new production line with an overall increase in the total amount of produced shoes.

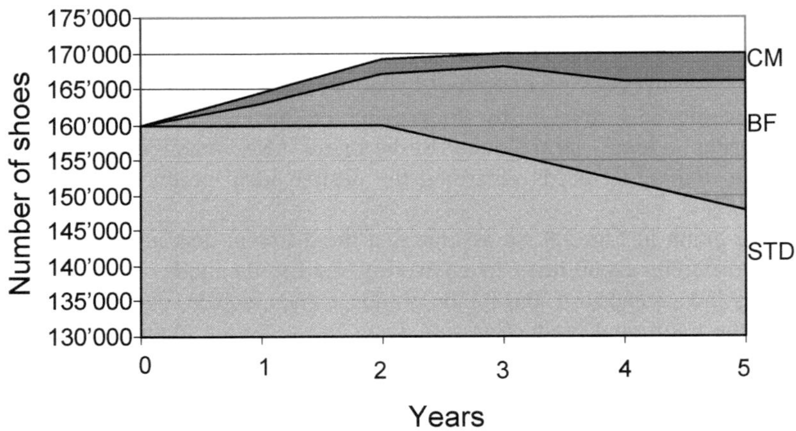

Figure 2.9. Example of demand level for scenario B

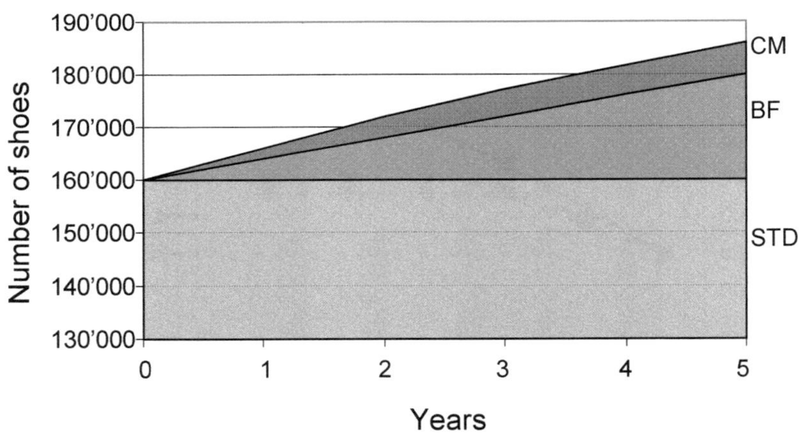

Figure 2.10. Example of demand level of scenario C and D

The business model includes the possibility to also tune the price of the MC shoes according to the product level chosen by the customer. In fact the shoe price is based on the combination of different components (colours, heel, outsoles and kind of leather) that can be changed/added according to the customer preferences. It is assumed that the price is a normal variable, with a mean value and variance which is different for BF and for CM (in the case of CM we consider higher variance).

Empirical research has shown that customers are willing to pay up to 20 - 30% above normal price for the increase of service they get from shoes that fit better to their needs than the standard shoes available on the market [18]. It is to be considered that with MC there is a shift in the purchasing decision which is less influenced by the level of price and more by the value of the product. The customer accepts to pay an average price higher than for standard shoes both for the intrinsic value of having a personalized shoe and for the experience of choosing his or her own features at the sale point.

2.7.5 Assessment of the Strategies

All scenarios are evaluated by forecasting the cash flows related to the adoption of new business models based on MC strategies. Investments are assessed first using the net present value (NPV) then applying risk analysis to evaluate the consistency of each scenario.

Some previous studies evaluated the economic relevance of MC in different sectors. A study based on the NPV calculation for the introduction of a laser sintering tool which offers the possibility to adapt any car design to specific customer requests was carried out in the automotive sector [19]. Kotha [20] examines the dynamics of implementing MC in a firm that pursues both mass production and MC in two different factories, making use of a detailed study of the National Industrial Bicycle Company of Japan. Reichwald etc. [21] evaluates MC from an economic perspective applying a general framework on a special setting of decentralized, customer centric production units (so-called mini-plants) located in close proximity to a particular local market in order to reduce distribution costs.

The analyses presented in this book calculate the best conditions in which to apply the MC paradigm in a footwear company according to its investment capabilities. It is assumed that the investment is monitored over a period of five years. In the computation model, cash flows are discounted using investment and inflation rates congruent with the cost of financing typical of the entire sector. Costs of production and similar assumed in this study are to be considered as incremental since we evaluated only extra costs and revenues linked to the application of MC strategies which are compared to standard production. The model was created considering data and information collected from:

- a traditional shoe producer for what concerns costs for production of standard shoes and the typical operational costs in traditional shoe factories
- simulations carried on forecasting the behaviour of traditional and innovative production systems with MC production in order to extrapolate data on production time, production capacity, etc. for the innovative production line.

The variability of the NPV of an investment project is an indicator of the project risk level. As well known in literature, considering only the payback period is too limiting because it considers only the short term view on the investments; considering only NPV means choosing between investments according to their monetary value; considering also the variability of NPV permits to choose according to an analysis of many different stochastic variables that influence the whole scenario. The evaluation of the variability of the NPV in this analysis is based on Monte Carlo simulation which generates a (pseudo)-random sample according to statistical behaviour of the most important variables , such as quantity, price and discount rate.

2.7.6 Presentation and Comment on Results

According to the collected data and their relationships (variable costs are linked to demand level and other direct and indirect costs driven by quantity and/or price), the above strategies for MC introduction give the possibility to increase the NPV with different impact according to the variability of the demand and correlated variables (costs, price, *etc.*).

The evaluation shows that the NPV of such investments for a traditional shoe producer is always above zero. If we compare the four strategies it is possible to see that its value grows since the increase of costs and investments is counterbalanced by the increase in the demand level which can be satisfied thanks to the introduction of new technologies (see Fig. 2.11).

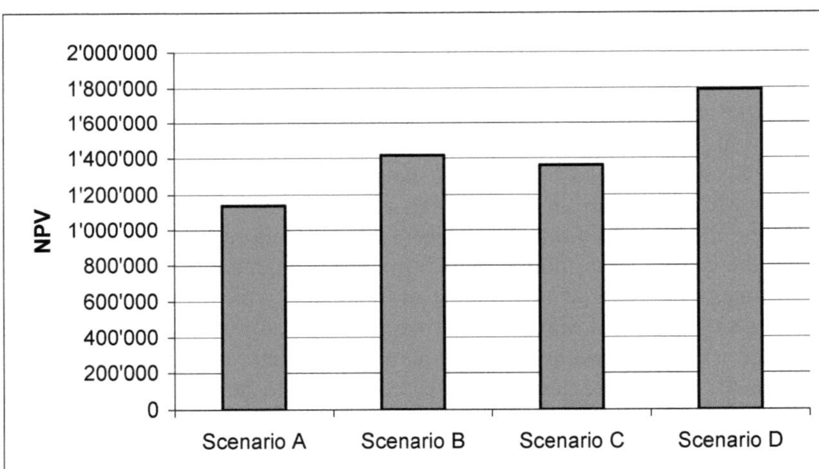

Figure 2.11. NPV for each scenario

In particular in scenario A the introduction of the scanning system for collecting information on the customer foot and the support received from a service provider for the feasibility study permits the company to sell a larger amount of shoes, even though some part of standard shoe sales is overtaken by the BF ones. With strategy B, the cost of machine for last production is compensated by cost saving on outsourcing of last production and by the increase in the level of satisfied demand with CM shoes. In strategy C the big investment in ICT and production technologies permits the company to increase the number of sold BF and CM shoes but the increase of NPV is limited due to large investments. Cost savings can be registered also in the percentage of initial stocks and in the amount of raw materials and components which the company manages to reduce thanks to a better organizational system.

Even though strategy D requires larger investments at the same time the total demand level that a shoe producer can satisfy is higher thanks to the greater flexibility of the system (achieved with the introduction of the logistics system) and the capability also of answering urgent orders. The NPV also benefits from an indirect impact on some other variables such as raw materials management or shoe delivery time which allow the realization of higher revenues.

In Fig. 2.12 it is possible to see that the payback period is very dependent on the type of investment undertaken: the first two strategies (A and B) requiring little investment can be quickly recovered while C and D require longer payback period (between two and three years). The investments in strategies C and D are more risky in the first year but in the following years they overtake the results of the other scenarios A and B.

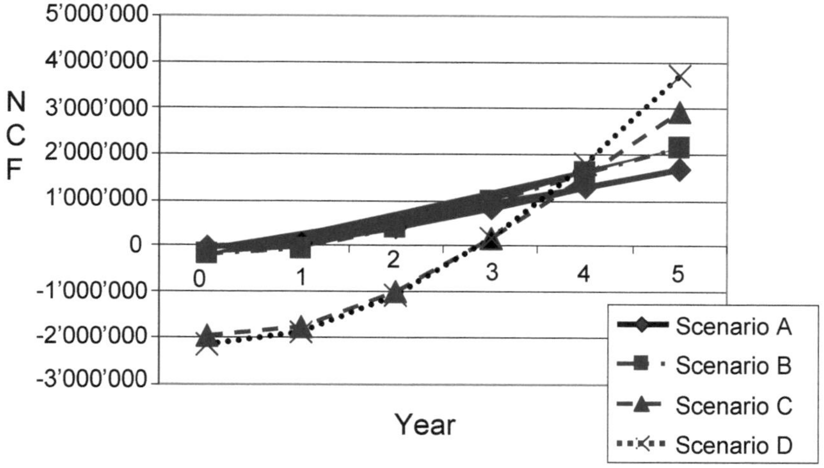

Figure 2.12. Cumulative net cash flow for each scenario

The goal of this work was not to define the best scenario for a shoe producer, but it is important to show how a company can decide to apply MC according to different strategies taking different degrees of risk.

2.7.7 Risk Analysis

After a static analysis, a Monte Carlo risk evaluation was carried out. Risk analysis assesses the same effects as a sensitivity analysis does, but also takes into account the probability distribution of the input variables. The four investment scenarios can be analyzed using this method because they are independent from each other, and a company can apply only one among the four without problems of stochastic dependency which can sometimes arise with Monte Carlo simulation. The most important variables, such as quantity, price and discount rates, have been given a probability distribution according to some hypotheses on their statistical behaviour based on historical data evaluation. For a matter of simplicity it was assumed that quantity of sold shoes has a triangular distribution while the price has a normal distribution with the hypothesis that over the years the price is either the same or does not decrease. Moreover it is assumed that the variability of MC shoe price is higher with respect to standard and BF shoes and it is justified by the fact that (as already mentioned) the company can apply a different price to almost each customer according to the level and the type of customization required. Variable direct and indirect costs, change according to the relationships existing with the quantity and prices. Results of the four scenarios are shown in Fig. 2.13 where it is possible to see the distribution of the NPV.

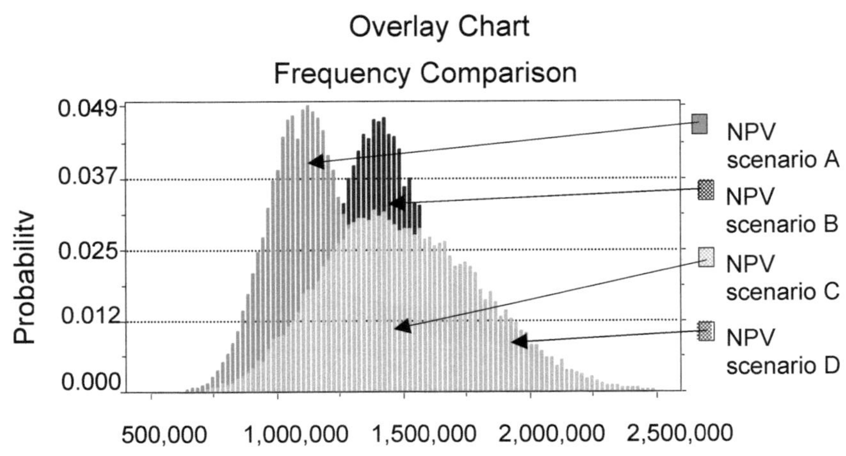

Figure 2.13. Comparison of the NPV distribution of the four scenarios

All the scenarios have a low risk level since the probability of having a negative NPV is zero. The strategies A and B have lower but more stable mean NPV values since their standard deviation is much lower than in the other two scenarios C and D. Higher variability in scenarios C and D is linked to the higher level of initial investment required, but they guarantee higher level returns.

The risk analysis permits to evaluate also which are the most important input variables influencing the NPV in each scenario, according to their behaviour and their correlation with the NPV distribution. As an example we show the case of the scenario A (see Fig. 2.14) where it is evident that unit costs of raw materials for BF shoes largely influence the behaviour of the NPV as well as the discount rate. The variability of quantity level of BF shoes has a positive influence and it is more or less the same along all the years.

From scenario B the unit cost for raw materials and components for CM and standard shoes assume more relevance on the variability of the NPV while in scenarios C and D the variability of the discount rate influences the overall NPV due to the larger variability of the cash flows (see Fig. 2.15 for scenario D). Costs of raw materials are generally a very influencing variable and in the case of customized goods they are even more strategic than in the case of standard shoes, given the higher variability that is to be managed. The just-in-time approach becomes even more a winning strategy thanks to the ICT software that can support integration between different providers and different departments.

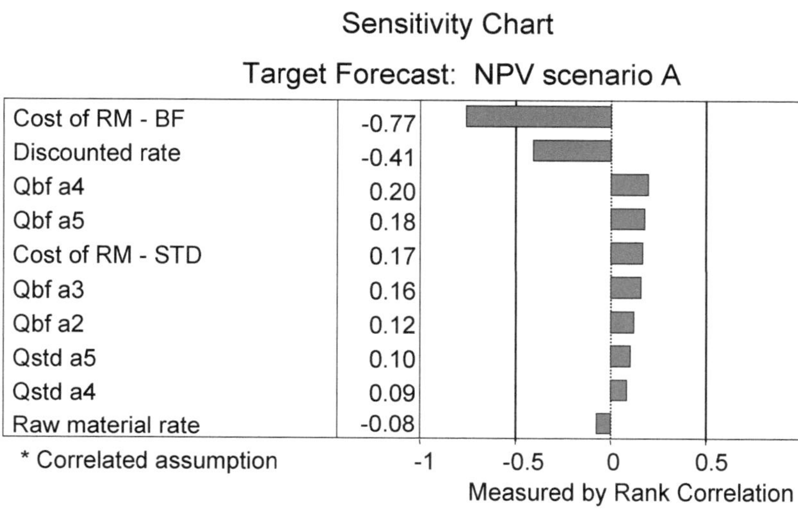

Figure 2.14. Tornado chart for scenario A

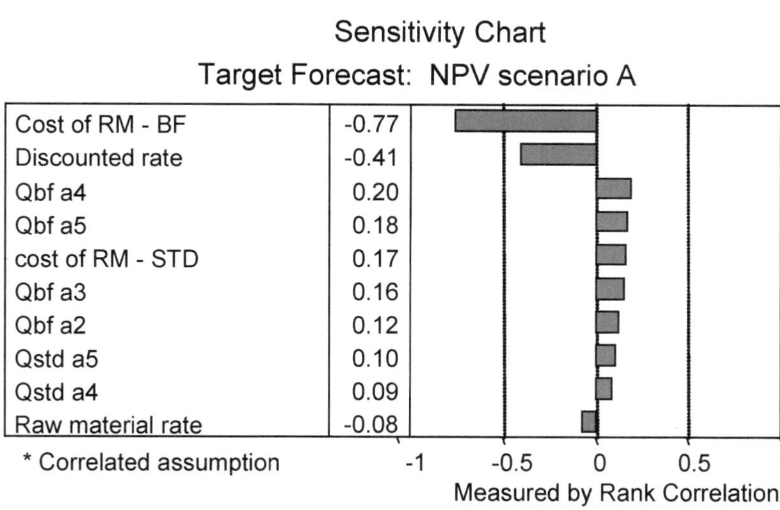

Figure 2.15. Tornado chart for scenario D

2.7.8 Further Qualitative Evaluation of MC

The model that was applied is based on data collected from the case of a modern strategy for implementing innovative services and production processes in a traditional shoe plant. The risk analysis analyzes how stable they are in order to understand whether it is a convenient approach or not. The limit of this analysis is in the fact that not all the incremental costs and benefits have been considered. From a late qualitative analysis what is still missing is a detailed assessment of some implicit benefits.

When applying an MC strategy for example, initial costs for marketing are certainly very high but the possibility given by the new tools to gather information on customer requirements directly from the final consumer permits a reduction in the "communication chain"; in fact most of the time a shoe producer has problems in establishing direct contacts with the final consumers because of many intermediaries (sales managers, dealers, shops and so on) and the only way to collect information on new fashion trends is to go to fairs, do market analysis, do brainstorming with designers and sales managers all of which take a great amount of time [18]. Having a database collecting all the data directly from customers makes it possible to ease statistical and market analysis, which reduces the time taken in the preparation of new shoe models (see also the considerations presented earlier).

The integration of the supply chain is another important benefit that is not fully quantified and forecasted in this analysis in terms of time saved and more efficient communication, and this factor is more relevant in scenarios C and D where ICT software is also implemented. Generally speaking, modifications of the organizational structure to facilitate mass customization, transforms the relationship structure and brings long-term benefits. Yet this is not to be seen as a great limit of the analysis, because this means that the NPV can be even higher for each scenario according to other potential benefits not considered here.

Given the impact of different degrees of innovativeness for a company willing to apply MC, it has once more proven the relevance of implementing innovative enabling technologies in manufacturing. The evaluation of the impact of new integrated ICT tools and new manufacturing technologies is extremely important in order to avoid under- or overestimating their impact on the company capability to produce and innovate.

MC represents a way to complete a product range adding new services for the customer in the view of the "extended product" provision. The applied model considers not only product related costs but also all the costs related to the implementation of the new services both for the customer (e.g. sales support) and for the producer (e.g. planning support). In such a context more than the mere economic data, it is important for a shoe producer to understand what steps to undertake as explained in the scenario description in order not to make mistakes and to avoid risky actions. As shown in this part of the work, this can only be achieved through the detailed definition of constraints and assumptions for each new scenario and evaluating all the variables concerned. In the model applied, investments are evaluated as adequate according to the degree of change a company is willing to apply. Most of the time the organizational aspects are the most difficult to consider and to re-engineer especially when we are talking about an SME. Further improvements to the model adopted should consider the whole life cycle of the shoe in order to evaluate the impact of technological changes not only for the shoe producer but also for all the other stakeholders of the pipeline (from designers, to suppliers, to customers, to dismantlers) and all the life-cycle phases of the product and process.

3

The EUROShoE Project

3.1 The History of the Project, its Goals and its Partners

The EUROShoE (www.euro-shoe.net) project was part of the European Commission Fifth Framework Program within the specific programme "Competitive and Sustainable Growth – Call Growth 2000", and was in response to the key action of innovative products, processes and organizations. The project was aimed at a dramatic renovation of the concept of the shoe as a product and of its production, based on the transformation of a mass produced product to a mass customized one. This product evolution goes in parallel with a transformation of the footwear company into an extended and agile enterprise [22]. Thus it needs to be capable of handling the complexity that such a change in the nature of the product implies and also needs to master the new challenges derived from a direct involvement of the consumer in the design and manufacturing process of the shoe. The problems that the project addressed were of great relevance for the shoe industry, which has been keen by monitoring them, thus allowing a short time to the exploitation of the results.

EUROShoE was a challenging and ambitious project aimed at an important transformation in the actual process of designing, producing and selling shoes. The project was aimed at paving the way for the development of a complete range of enabling technologies that, with or without the aspect of the product customization would have set new standards in terms of flexibility, process efficiency and technical levels for basically all kinds of shoe production. It was also a test bed for integrated projects (IP) in the Sixth Framework Programme of the European Commission, having visionary goals, ambitious results, with a large number of partners encompassing the whole product and process life cycle.

In order to tackle such a complex task, a consortium of 34 European partners was formed. This was thought to be necessary for two distinct reasons, first, the knowhow deemed necessary to achieve the research goals could not be found in one single country, and had to be gathered from private companies, universities and research centres, with both general and specific expertise. The second reason was related to the aim of developing solutions with a general validity for shoemaking companies from all the major European footwear-producing countries. The results of the project contributed significantly to a general improvement of the sector. For this reason the contribution of shoemakers from different countries was vital in its success.

Such a radical change in the product nature forced a complete revision of the processes that support the various phases of the product life cycle, namely design, production, sales and distribution, use, dismissal and recycling as indicated by the model ManuFuturing [23]. Figure 3.1 is a systemic view developed within the EUROShoE project according to the model of the product–processes matrix. This has led to a research effort that encompassed the development of new technologies for each phase of the life cycle. This total and global rethinking of the footwear business needed large resources and the EUROShoE project has therefore been an ambitious and large research initiative involving all the actors in the value added chain.

The project took two major viewpoints for the research, first, exploring the 'shoe system' requirements and, second, developing the necessary methodologies and technologies needed at every step of the value added chain. Moreover the entire project aimed at a Web-orientated transformation of the business and operational mechanisms of traditional enterprises such as footwear companies. This enables a typical "old economy" manufacturing business to evolve into a "new economy" business, with a transformation within its manufacturing base. The above transformation requires action in a number of interlinked areas:

- definition of a 'reference model' of the business and its operational processes [23]
- development of a set of software tools and procedures to select, configure and integrate enterprise resource planning (ERP), product data management (PDM), and computer-aided design and manufacturing (CAD/CAM) systems [24]
- implementation of the full ERP/PDM/CAD/CAM environment [16]
- design and development of a new generation of foot-feature capturing devices and camera-based foot scanners
- creation of a knowledge-based CAD/CAM software for the design of customised shoes [25]
- development of versatile, multi-purpose shoe machines and systems [26]
- realization of the components of physical and virtual (Web-based) sales centres, where customers' foot measurements can be carried out, styles selected, data transmitted to the manufacturers and transactions conducted [27, 28].

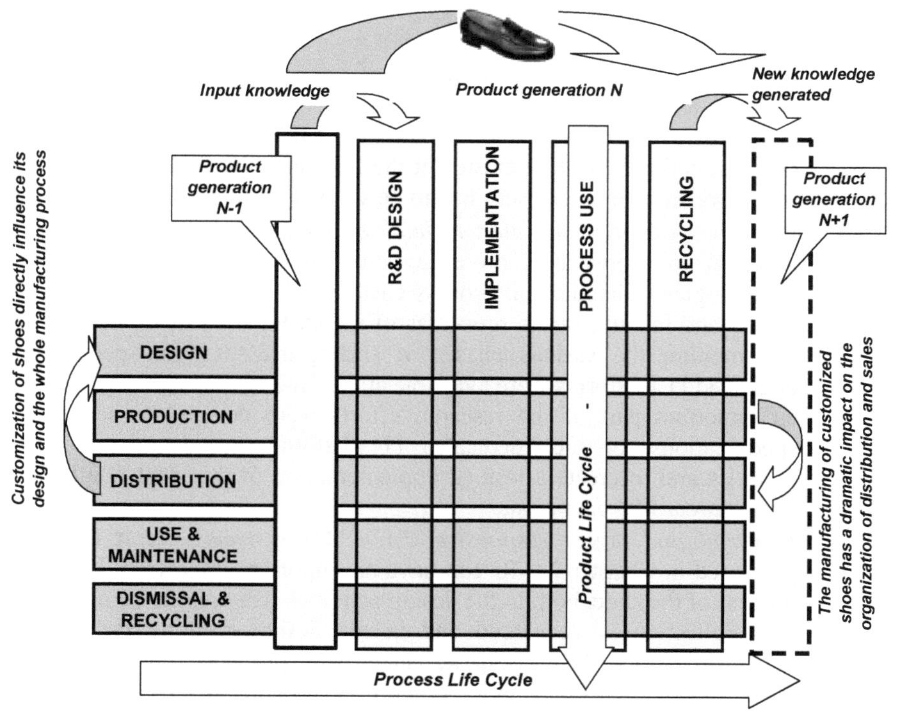

Figure 3.1. Product - process life cycle matrix

3.2 The Technical Challenges and the Main Innovations

The EUROShoE project was the first structured and comprehensive research initiative aimed at exploring the application of the mass customization paradigm in the footwear business; its prime goal was then to identify the key enabling technologies that would have been necessary for this implementation, to assess their state of the art and, whenever necessary, develop the missing ones.

In order to achieve its goals, the project had to face the following technical challenges:

- *Adequate tools for capturing foot features* - the first step to be performed in order to start the process of designing and manufacturing customized shoes is the measurement of the consumers' feet. Therefore foot scanners are key components of the set of enabling technologies needed to support the operations of the enterprise. Although some basic solutions were on the market at the beginning of the project, none of them fully responded to the specifications of the EUROShoE framework. Therefore a new family of advanced devices aiming at capturing the features of the foot had to be developed.[22]

- *Matching and tailoring of the last* - software applications to match foot and lasts and, in case the matching is not satisfactory, to modify the last in order to create a tailor made one, are also critical components of the sale process; nothing like that existed at the launch of the project, so efficient and powerful computer codes had to be studied for this purpose.
- *Extensive use of information technology* - mass customization, and its adoption to footwear is no exception, relies heavily on information technologies to handle and process customer data in order to initiate the design and manufacturing processes of the customized shoes and to plan and monitor the various phases of such processes. Non-specific and dedicated IT platforms were available at the time of starting the project, so an important part of the research efforts were devoted to setting the specifications for such dedicated IT platforms, to identify the basic modules and to create a testbed implementation of the ideal IT platform for footwear MC operations.
- *Powerful and highly responsive CAD/CAM systems* - as it has been presented in Chapter 2, MC can have an important impact on the design process of the shoe, both in the design activities that are carried on when a new collection is developed and in the design tailoring task that is performed when fully bespoke shoes are produced. In both cases CAD/CAM systems play an important role and they must have high levels of performance and responsiveness. The EUROShoE took as a background two well known and consolidated shoe CAD/CAM software packages (Shoemaster© from CSM3D and PowerSHAPE © from Delcam) and developed on top of them a whole new family of dedicated functionalities to support in the most effective way the design side of the MC business.
- *Flexible and highly automated manufacturing facilities* - as for the design tools, also at the shopfloor level (manufacturing process) the machines that perform the different production steps must be highly flexible, easily programmable and very much automated; however not all the shoe machinery on the market at the time of launching the project had an adequate functionalities to fulfil that level of expectations. So some machines had to be rethought and improved, and for others a new approach to performing their relevant tasks by using robots was developed.
- *Pervasive planning and monitoring of the key processes* - last but not least, MC operations, due to their high level of complexity and the many challenges posed by the responsiveness, cost and time constraints, need to be carefully and effectively planned and scheduled at all levels and properly monitored at their execution stage. Again, specific applications were lacking or not enough knowledge was available on the applicability

[22] Some of them later evolved into industrial products and have been included in Section 2.6.

of the existing ones to the specific context of the footwear MC business. Also in this area the EUROShoE project pioneered methodologies, software applications and tools whose applicability was then demonstrated in the final validation phases of the project itself.

In the three years of its duration, the various research teams involved, managed to respond to these challenges with a list of "technological proposals" that led to the realization of a number of prototypes that, in the course of the project, were extensively tested in a quasi-productive environment.

3.2.1 Innovations in the Design Process

As indicated in Chapter 2, two of the five main phases that constitute the overall process of offering consumers customized shoes involve design and its related activities. It should also be clear that the implications (constraints) on design deriving from involving the consumer in the sale-design-manufacture-distribution loop depend very much on the degree of customization adopted. This explains why a great deal of attention in the EUROShoE project was given to developing new CAD/CAM tools specifically suited for the mass customization application.

In the best-matched fit case, design activities are performed in a way that is very much the same as in the traditional shoe business; the costumer is involved only at the sales level and the "big" change (for the shoe maker) is the move from a made to stock to a made to order approach (sales come first and then manufacturing). In addition, the design procedures, followed from style concepts to samples, are the usual ones. Yet a certain influence of having the individual consumer taking part in the value chain of the product is likely to be there: if it is planned to allow him or her to "configure" the product the way he or she likes and to choose the last that really matches closely the dimension of his or her foot, then this aspect must be taken into consideration in designing the new collection of styles that will be offered as "customizable" models.

In fact giving the consumer the possibility of "configuring" the product means that all the possible configurations (each one identifying a precise combination of all selected parameters such as upper material and/or colour, lining material and/or colour, bottom material, heel height, accessories) must be managed both at the level of design and, in particular, of product data sheets and bill of materials/components. The same problem appears also in relation to last size (length) and fit (width at the ball girth) combinations. In fact the closer the dimensional match of the last to the feet biometric data of the consumer, the better the customized shoes will fit; this could mean handling up to 30 sizes (full numbers and half numbers) within 3 fits per each size which would result in 90 last geometries for each individual product (in the worst case scenario). And the CAD tool that is used to support the design activity must be able to handle such level of complexity, to model, grade and flatten all the different size–fit combinations. This is quite challenging for even the most powerful and complete CAD tools available on the market; in the EUROShoE project the Shoemaster© software had its functionalities consolidated and enhanced in order to fulfill the above mentioned specific requirements (Fig. 3.2).

Furthermore the attention to biomechanics introduces some additional challenges and complications: not only must "biometric" aspects[23] be taken into consideration, but also those other factors in the shoe design that might affect the dynamic motorial performance of the body when the consumer wears the new pair of shoes. These biomechanical aspects have usually been considered by shoe companies in developing their models, although the degree with which they actually affect the design vary with the typology of the shoe itself; companies involved in sport shoe production have always dedicated a great attention to these parameters, being very much aware of the importance that a correctly made pair of shoes can have on the performance of an athlete; also non-professional users of sport shoes are starting to realize the importance of such design factors on the comfort and efficiency of their footwear.

The orthopedic sector is the other classical example of how the design of the shoe must incorporate a thorough understanding of the static and dynamic effects of the footwear on the human body. The notion of dynamic effect is also associated with the notion of customization, each pair of orthopedic shoe being strictly tailor made for each individual patient. So design constraints dictated by biomechanics do not necessarily relate only to customization, but they could rather bring an "added value" to the design of the shoe; the EUROShoE innovative approach to that statement was in allowing these constraints to become part of the normal design flow of the shoe models, which required an understanding of the basic biomechanical phenomena, their translation into correlations between a desired dynamic performance and a given design parameter, and, as a final step, the formalisation of such correlations into "rules" that the CAD software can interpret and use to drive and support the shoe engineer in his design work.

In order to achieve that, a new ingredient in the recipe of the design tools was introduced: the knowledge base (KB). This can arguably be regarded as a leading edge advancement in the evolution of shoe CAD systems; the integration of a KB actually gives the possibility of "capturing" the design experience (through the formalization of design rules, not only related to biomechanics, but also of a more general nature or company specific) and to make it available to the designers who use the system. An entire work package in the EUROShoE project dealt with the exploration of the possibility of integrating a KB, filled up with the appropriate knowledge, with a typical shoe CAD software. The knowledge the CAD relied upon was generated in an extensive experimental campaign conducted by two major biomechanical research institutions in Europe to identify the most important relationships between static parameters of the foot and shoes, their influence on fit and comfort, or the interaction between last and shoe features and the motor performance of the individual wearing the shoe.

If that represented the real leading edge development in the area of CAD (the key enabling technology to support the design process), some less "visible" but not less important work was performed with the aim of simplifying and speeding up the design work; the time factor is particularly relevant when the design tailoring

[23] These are the aspects that refer to the morphology of the last in relation to that of the foot and of its characteristic dimensions as well as to the characteristics of the shoe made on such last.

(adaptation of the basic shoe proportion to a tailor made last for a pair of truly bespoke shoes) activity must be performed as part of the daily process of design and manufacturing of custom fit shoes (in EUROShoE terminology).

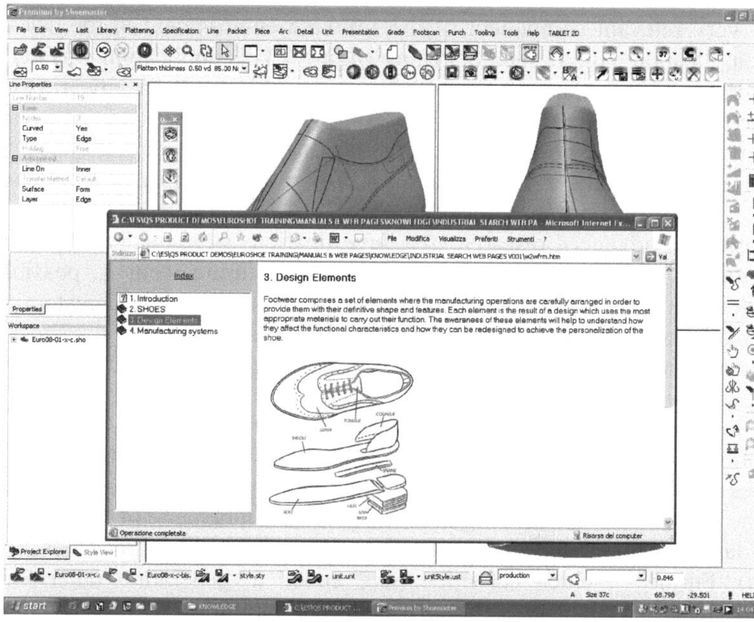

Figure 3.2. The CAD package Shoemaster©; improved version used by the EUROShoE project (courtesy of CSM3D)

This activity technically implies transferring all the design features (style lines) of the basic shoe model chosen by the consumer from the original standard last to the modified tailored one. To accomplish this task, specific CAD functions (transfer functions) are used, which normally imply high levels human intervention, attention and skill. The work done in the EUROShoE project, was to make these steps more automated, simpler and faster so that the time needed to perform them was compatible with the time constraints of the mass customization environment.

3.2.2 The Innovations in the Sale Process

Quite an intensive research effort was dedicated to the development of a new generation of foot feature capturing devices (to differentiate them from the foot scanners already available at the time), with specific explorations in two different directions: multi-sensorial approaches to acquire at the same time dimensional information of the foot as well as biomechanical data; and low cost and portable scanning technologies to simplify the operations in low investment sales points (with the possibility in the future to also extend this technology to Web-based scanning). Scanning devices are an important ingredient, but they are not the only ingredient that supports the sales operations of the customized shoes: an efficient product configurator (a software that can run on different hardware platforms), a very powerful consumer profiling system and an efficient data exchange and communication platform are also vital parts of the solutions that were studied in the project.

One very relevant component of the software environment at the sales point was an application called "shoe matching engine". This application was dedicated to acquiring all relevant profiling information related to a given consumer (biometric data, biomechanical patterns, habits, preferences and so on) in order to use them to drive the consumer in selecting the most appropriate shoe and to identify (or generate the data to create a new one) the last that best matched this profile.

As a further advanced research area, investigations were made in the direction of introducing new sensorial cues that could compensate what is, possibly, the most evident drawback (for the consumer at least) when purchasing customized (or better, made to order) shoes: the fact that the customer cannot test the shoes in the shop and cannot evaluate them personally and immediately from the aesthetic and comfort point o view.

An augmented reality approach was explored, which combined visual information from the real world (a video of the consumer) with digital images of the shoes selected and configured so to give an image of the consumer wearing them (the device that was investigated was given the nickname of "magic mirror"; it was later patented and further developed and improved in other research projects). Further research activities were aimed at generating other kinds of sensorial cues that could give the consumers a "comfort feedback" even if they don't wear the real shoes. These two very advanced research topics proved the feasibility of such concepts.

3.2.3 The Innovations in the Manufacturing Process

Other important research chapters were focussed on manufacturing, for which a whole range of new machines and systems were developed, all complying with the basic requirement of the utmost flexibility and versatility (which was translated into the motto: "*zero tools and zero set up time*"). The project work in this area went from enhancing cutting operations, to last milling, to toe lasting, to robot based roughing, cementing and finishing operations, to handling materials, components and the work in progress in the making room by means of advanced automated transportation and logistic systems; a great deal of work was dedicated to computer planning and monitoring of each individual shop floor operation. Providing a detailed list of all the numerous innovations studied in this area is out of the scope (and of the space available) of this book, but the following short-list gives an idea of the many parallel investigations preformed in the EUROShoE project:

- A new architecture of a dieless cutting machine with an innovative conveyor belt system to move the materials to be cut through the machine stations and equipped with an automatic pick device to collect the cut parts.

- A super flexible toe lasting machine[24] equipped with "variable geometry" tooling and fixtures to minimize tool changes and set up times.
- A fully integrated, automated, software controlled logistic system aimed at storing and dispatching all relevant shoe components within the factory.
- A robot based roughing and cementing cell (typical shoe making operations) based on an innovative "active" system to monitor and control the roughing force applied in order to always achieve the best processing quality.
- A robot-based finishing cell, which, for the first time in the sector, explored the possibility of using robots in the finishing operations of the shoe[25], normally considered to be very demanding activities for which high levels of human skills are needed.
- An innovative NC last milling system in which the last (standard or tailor made) could be milled from CAD data in one single combined operation and on one single machine.

Last, but not least, since mass customization is more than anything else handling and mastering digital information, a great research effort was dedicated to analysing the organization and the operational mechanisms of the footwear industries (a work that was performed using advanced enterprise modeling techniques) to highlight the changes that were necessary, from the organizational and information system standpoint, to migrate to the new business approach. This very intensive methodological activity was aimed at the selection and implementation of the complete information system's platform (which included software modules typical of an ERP environment, a PDM system and a Web portal to interface the network of suppliers) to handle, plan and supervise the operations of the entire enterprise.

3.2.4 The Integrated Pilot Plant

Probably the most relevant legacy and the most important intrinsic value of the project, was the so-called Integrated Pilot Plant (IPP). Besides being beneficial to the research teams involved as a common and integrated test facility for the final tuning and development of the various machines and equipments, the IPP represented a unique implementation that had (at the time of the project) no equals anywhere. The IPP was considered as a "model shoe factory'" highly integrated and with very high levels of automation in all its processes; in the final configuration the plant was able to produce customized shoes with various degrees

---

[24] The toe lasting machine is used in shoe making to assemble, on the last, the front part of the shoe (the toe region) with the insole (that is the structural component around which the shoe is made); toe lasting machines are very important in a modern shoe factory and normally the quality of this operation determines the final quality of the shoe.

[25] These are the operations with which the shoe is cleaned, brushed, waxed and polished in order to remove the dirt and stains accumulated during the manufacturing process and to give it its final glossy and elegant appearance.

of customization as part of the validation phase of the EUROShoE project, with a twofold objective:

- To test the family of enabling technologies conceived to support all the processes of the life cycle of customized shoes (design, sale, design tailoring, and manufacturing).
- To prove that the developed technologies could actually provide the desired technical and economical results by running an extensive validation campaign in which all such technologies were deployed and operated in a way as similar as possible to that of a real shoe company.

Therefore the IPP had a crucial role in demonstrating how far and how much the different research chapters (and the related software and hardware prototypes) had achieved their goals. Although not at the same level of productivity achieved in the final phase of the EUROShoE project, the IPP is still operational in the research laboratory that ITIA–CNR has in Vigevano; it consists of two main sections:

- the control deck
- the shopfloor.

The Control Deck - the control deck is the equivalent of the sale, administration (partially), production planning and design departments of a shoe factory. The functions performed at the control deck are as follows[26]:

- **Order processing** - orders coming from the local or remote sale points differentiated by shoe model and level of customization are gathered by this application, sorted and prepared for the subsequent scheduling.
- **Production order processing** - sale orders are processed at this point to generate production orders, initially at the level of master production schedules and then on a more refined level. Material requirement planning is also performed at this stage.
- **Finite capacity scheduling** - at this stage detailed finite capacity schedules, based on the MPS, are generated department by department at the factory level (high level finite capacity scheduling).
- **Production lot processing** - this stage generates the correct sequence of production lots based on the customer orders and taking into account constraints and requirements of shopfloor logistics.
- **Supply chain management** - this function takes care of the scheduling and monitoring those production phases that are outsourced to external suppliers and it ensures a direct control on their execution.
- **PDM repository** - the product data management environment is used as a repository for CAD/CAM data also accessible from outside the company; as in the case of the external design office that performs the design adaptations to be carried out in the case of fully custom-made shoes.

---

[26] Most of the applications listed were off the shelf software applications that were selected as part of the research project, customized and implemented at the IPP.

- **CAD** - this is the design environment of the IPP in which the models belonging to the EUROShoE collection were initially designed and engineered. Full 3D CAD files of all the models and variants are managed by this software tool. Bill of materials and manufacturing sequences are also defined at this stage, although the creation of such data and their management is taken care by the ERP system.

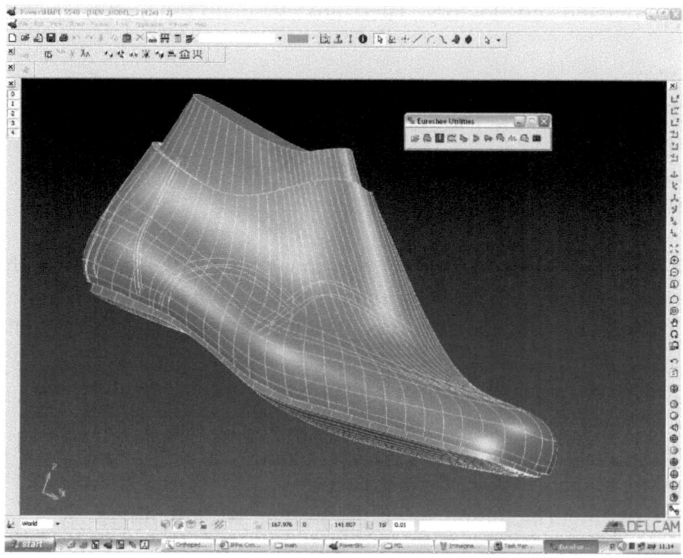

Figure 3.3. The CAM environment of the EUROShoE project (courtesy of Delcam)

- **CAM** -this function takes care of the generation; starting from the 3D CAD data, of all the relevant information, both for planar or space processing, that will be used by the automatic machines that are present at the shopfloor level. A command toolbar is used to progress step by step through the various phases of CAM processing and ensures that all the necessary tool paths are created and correctly stored in the PDM environment (Fig. 3.3).
- **MES** - the manufacturing execution system acts as a workflow management system handling and distributing information between the CAD/CAM/ERP environment and the different machines and applications at the shopfloor level.

The applications installed in the control deck form the IT platform of the IPP. They support the functions and processes that occur in the various phases of the creation of a new shoe collection (design) up to the complete engineering of the new models in terms of CAD data, material and manufacturing details, and are also used in the phase of the generation of consumer specific variants of the models of the collection in the so-called design tailoring phase. In such a scenario all the described applications and the corresponding phases of the process are repeated for each single customer order. A broadband connection to the Internet completes the IT infrastructure of the control deck; through such a connection high volume data can be exchanged between the IPP and its remote partners (both on the design and manufacturing side).

The shopfloor - The shopfloor level of the IPP is organized as a fully integrated and highly automated shoe factory. There are five main functions/departments:

- cutting
- stitching
- internal logistics
- last making
- making/finishing

Cutting department - This department is responsible for the very first phase of the shoe manufacturing process: cutting out all the materials that will form the shoe. The cutting department of the IPP cuts all the leather and synthetic materials that make up the upper (outer part of the shoe), the lining (inner part of the shoe) and all the other non-visible components. The operations performed include:

- leather inspection
- cutting and collecting the cut pieces
- pre-stitching (including splitting, skiving, NC stitching).

Leather inspection can be performed, in some cases, on a dedicated station where the outer contour of the hide is captured with a high resolution digital camera, while quality areas and localized defects on the skin are manually marked and classified by the operator. The digital map of the hide with all the relevant quality areas and defects is then made available to a dedicated nesting software that computes, based on the manufacturing order specifications, the most efficient layout of pieces to be cut. The hide inspection station is connected to the shopfloor Ethernet broadband network.

Cutting is done with the new generation dieless cutting machine developed in the project (Fig. 3.4); a high frequency oscillating knife ensures maximum speed and maximum cutting quality. The machine has the capability of both processing pre-nested hides, which are simply loaded onto the machine, or of nesting locally and interactively the patterns to be cut in the machine itself. In such a case a projection system projects the shape of the patterns to be cut on the loaded hide and the operator can then position it in the optimal location. The system is equipped with a conveyor belt that, while the cutting process progresses, moves the hide to the unloading section of the machine.

The pattern collecting phase is taken care by a second device based on a programmable and reconfigurable prehensor that is instructed by the control software to move to the location of the pattern to be picked up and to configure its fingers so to adapt to its shape. The collected patterns are automatically placed into a box at the very far end of the cutting table. Boxes are then coded and associated to the specific customer order they belong to.

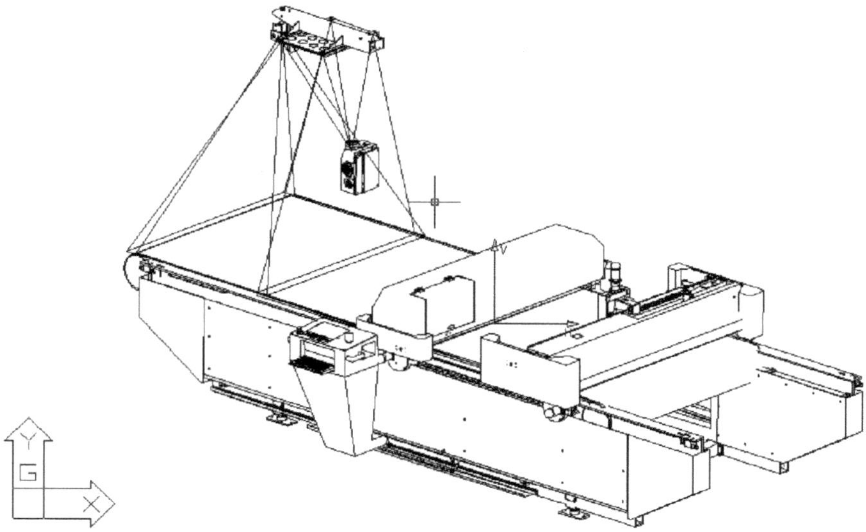

Figure 3.4. The dieless cutting table (courtesy of ATOM)

The cutting table is also connected to the shopfloor Ethernet broadband network. Boxes are then moved to the pre-stitching section where they are split (on a traditional machine) and skived.

Skiving is performed with a new generation skiving machine[27], equipped with servo actuators that allow a direct drive of the part to be processed with or without the limited intervention of the operators. An NC (numerical controlled) stitching machine is also present for the assembly of complicated parts and the machine is equipped with a pallet cutting station. As a final step patterns are stamped and labelled and then the box with the kit of cut parts is taken care of by the logistics system and loaded into the box warehouse. Components are then ready to proceed to the next phase, stitching.

Stitching department - The stitching department is based on an innovative highly flexible conveyor system. The conveyor system (fully automated and PLC controlled) serves four work posts (folding, planar stitching, back stitching and special stitching); boxes carrying the pieces to be processed can be moved by the system from one post to another one in a dynamic and flexible way, depending on the manufacturing (stitching) sequence each shoe has to undergo and on the line balancing. The stitching conveyor is supervised and controlled by a dedicated software application and is also connected to the shopfloor Ethernet broadband network. The conveyor is mechanically connected to the box warehouse from

[27] Because of its nature as a test and development facility, the IPP frequently hosted prototypes of equipment developed in other research projects; for example the skiving and folding machines and conveyor system of the stitching department, that were studied in the CRAFT project named CICLOP (computer integrated closing operations).

where boxes carrying the pieces to be stitched are moved to the stitching conveyor and where they are automatically sent back at the end of this phase.

The stitching department has its own low level scheduling capability, provided by a special implementation of a dedicated software application capable of performing a detailed department level scheduling of the jobs to be performed on the various stations. The departmental scheduler receives data from the factory planner (see description of the control deck) and uses them to produce the detailed level schedule for which for the department is responsible.

The stitching phase is completed by post-stitching operations, which include back moulding of the upper, toe moulding and eyelet application. These steps are performed on off the shelf standard machines; the boxes to be processed are called from the logistic systems, delivered to the relevant areas of the factory, and sent back again to the warehouse at the end of the post-stitching operations. Uppers are then ready to be sent to the making operations.

Internal logistics - The internal logistic system (Fig. 3.5) performs a two-fold function: storage for shoe components processed internally at the IPP and externally by suppliers; and dispatching these components to different locations, both in the stitching and in the making departments, at the right time. The system uses an overhead box storage rack with a pick and place automatic device that handles storing and retrieving operations of the boxes; four active towers with elevators are used to move the boxes from the shopfloor level to the storage level (approximately 2.5 m high). Conveyor belts are used to dispatch the boxes between the towers. The four towers have the following functions

- Component association/stitching tower - this dual access tower is used to receive components from external suppliers, to associate them to a specific customer order, and to deliver and retrieve boxes from the stitching room.
- Processing tower - this second tower has three distinct functions: to move boxes for the first time to the box storage after the pre-stitching operations, to deliver the boxes to the post-stitching operations and to move them back to the storage at the end of this step, and similarly to deliver/retrieve the boxes to the finishing operations.
- Lasting tower - this is used to deliver uppers and insoles to the lasting island of the making line to initiate the making phase.
- Sole application tower - this is used to deliver the box containing the sole to the sole application station.

All towers are equipped with a dedicated PCs; these four PCs are coordinated and managed by a central controlling station; they form a network that is then connected to the shopfloor Ethernet broadband network.

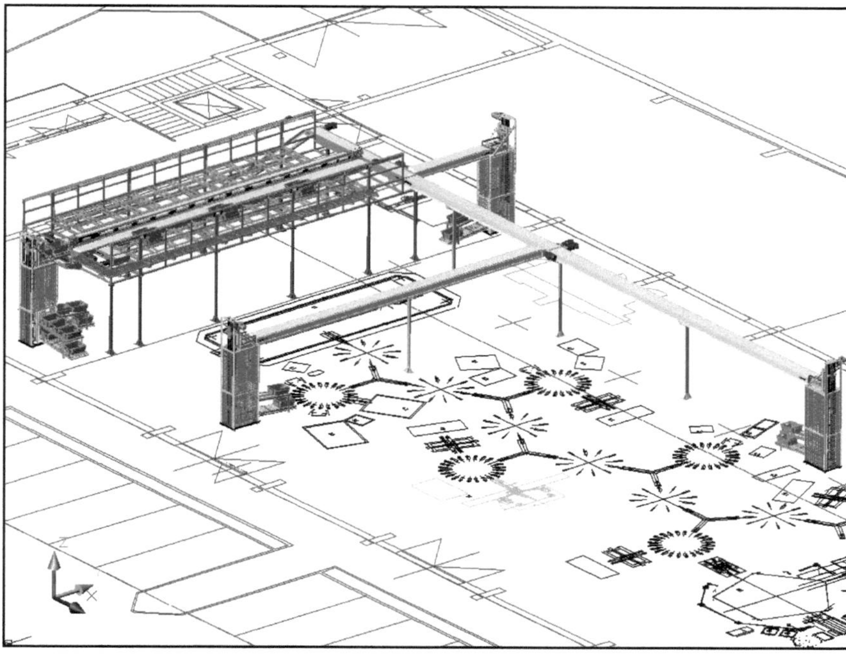

Figure 3.5. The integrated logistic system (CAD rendering) (Courtesy of LIREL)

Last making department - This is the department where lasts are manufactured, both for the so-called "best matched fit" shoes and for the "custom fit" shoes; it is based on an NC machining centre in which all the main steps of last making are performed: roughing, finishing, processing the toe and heel regions of the last, machining the top plane and the seat of the clamp, and drilling the holes for the mounting of the clamp. The machine processes pre-hinged blanks on which all the mentioned operations are performed automatically in sequence.

The last making machine is connected to the shopfloor Ethernet broadband network, through which manufacturing orders and geometries of the lasts to be milled are received. Such geometries are then processed locally to generate the tool paths to be sent to the numerical control. Lasts are then handed on to the last loading station, where the clamping block with the RFID tag is mounted; at this stage tag code and last code are associated and from that point the last is visible to the control system of the making line. Lasts are then stored in the last warehouse.

Making/finishing department - The making department is equipped with a fully automatic conveyor system, based on an architecture called "molecular" (after the resemblance that the layout of the making line has with the scheme of a molecule). The line incorporates a fully automatic last warehouse that can take up to 200 lasts. The conveyor system is controlled by a PC which supervises all the functionalities of the line from calculating the optimal routing of the shoes from one station to the other (depending on their manufacturing sequence), to handling the traffic in the line avoiding jams and bottlenecks; to supervising the duration of manual and automatic operations; down to controlling single actuators. Lasts (with shoes) in the line can be flexibly routed to any location, they can be buffered, they can by pass operations, and they can be returned to previous steps or back to the last warehouse. The conveyor line serves six processing 'islands' with the following functions:

- *Lasting island* - this is where toe and heel/side lasting operations are performed. The last first receives the insoles (fixed with no nails), then the upper is hooded on the last, and then, after humidification, the upper is lasted on the last. The toe lasting machine is a special prototype of the "super flexible" machine, capable of adapting to a large variety of toe shapes without the need to change fixtures (wipers and toe band). It is connected via field bus to the conveyor system controller for the exchange of the synchronization signals and via Ethernet to the shopfloor network to receive the tool path information to perform its job. Such programs are generated in the CAM step (see control deck). The heel and side lasting machine is similarly connected both to the field bus and to the shopfloor network and it is also mechanically integrated with the conveyor line so that it can receive and send back parts to be processed autonomously.
- *Bottom preparation island* - this island hosts the first thermal conditioning unit of the line (heat setter). Uppers are exposed to a warm stream of humid air to allow the upper to adapt to the shape of the last and to assume its desired conformation. The unit is operated automatically by the conveyor line and has a robotized pick and place arm to insert and remove lasts; traditional bottom pounding and heel setting machines complete the configuration of this island.
- *Roughing/cementing island* - a first ABB robot is installed here. The robot-based cell performs bottom and side roughing operations and bottom and side cementing. Roughing operations are performed with a dual purpose and dual tool head, equipped with a force control active electronic that has the task of controlling and keeping constant the roughing force. This is aimed at compensating irregularities of the upper and at maintaining a constant roughing quality all along the contour. The cell does not require any direct machine programming since programs are directly generated by the CAM software and downloaded to the cell at the right point in time; the cell is also connected to the shopfloor Ethernet broadband network. Synchronization signals are exchanged with the line controller via field bus. A solvent drying station, fully automated as in the case of the heat setter, is used to help the solvent of the glue to evaporate after its application.
- *Sole application station* - this is the station where the sole and the upper are assembled. For normal cemented shoes, soles and uppers, after reactivation of the cement, are pressed together to perform the bonding of the two components; two different kinds of sole pressing machines (an hydraulic and a pneumatic one) are used for normal soles and for walled plastic soles. For Blake stitched soles a special sole stitching machine is present.

- *Injection moulding island* - this portion of the line hosts an injection moulding carousel, which is used to produce shoes with direct injected plastic (PU) soles. The carousel is integrated with the line by means of a robotized pick and place arm which is used to load the shoes to be injected in the carousel and to retrieve the finished shoes after the injection process. The operations of the carousel are synchronized with those of the entire line via a field bus connection. A chiller (refrigerating unit where shoes are exposed to a stream of cold air) is also positioned in this island, its operations are fully automatic and controlled by the line supervisor as in the case of the two other conditioning units.
- *Finishing island* - some finishing operations are performed here. This island features a second robotized cell: another ABB robot (Fig. 3.6) is used to perform on the finished shoe a set of polishing operations that include: cream application; spraying cabin; wax application; multiple steps of brushing; and buffering between the various steps. The polishing sequence is controlled by a "recipe" that is specific for each shoe. The cell is synchronized with the operations of the line via a field bus connection; it is also connected to the shop floor Ethernet broadband network from which finishing programs created by the CAM applications are received. At the end of the manufacturing sequence an automatic last pulling machine pulls the last off the shoe and deposits the shoe in boxes at the end of the process. Lasts are sent back to start a new cycle.

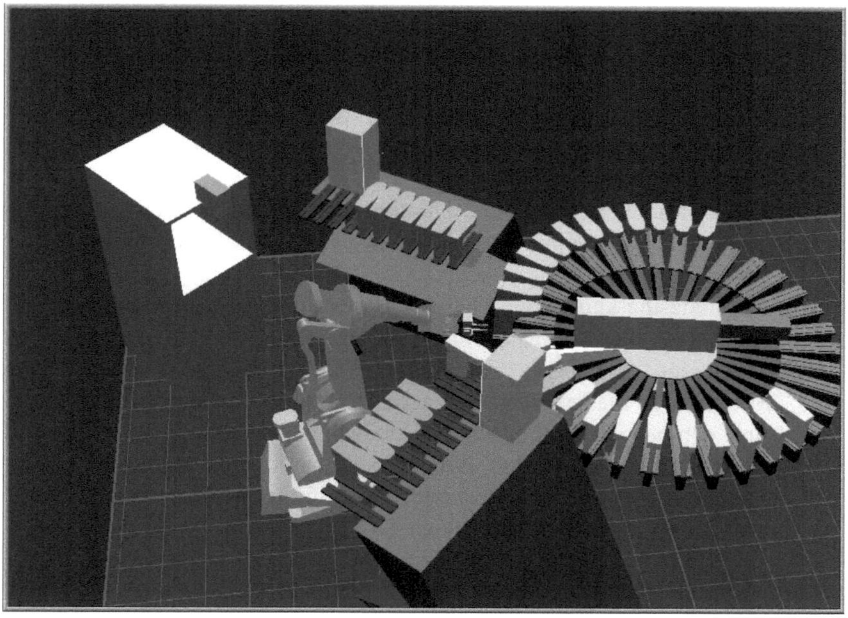

Figure 3.6. CAD model of the robotized finishing cell (courtesy Joseph Stephan Institute)

As in the case of the stitching department, the making department also has its own low level scheduling capability: a finite capacity departmental dynamic scheduler is available that takes care of defining the timing of the various operations performed in the corresponding work stations. The scheduler has the capability to dynamically reschedule the work in case of disturbances, failures, urgent orders and similar.

Soles are delivered to the plant pre-treated; an automatic sole cementing machine is used to apply cement to soles, both to normal or walled ones. Blake stitched soles require that the shoe is pulled off the last, the sole is then stitched to the insole and upper, and then the last is inserted back in the shoe.

The finishing department is mainly based on traditional machines; at the end of the making line each individual shoe is labelled with a label carrying the order/customer number. This label is used to recall the box containing the finishing components and to then start the final finishing phases, which include:

- upper trimming
- heel application and fixing on women's shoes
- steaming/ironing
- final brushing
- application of the cleaning insock and/or customized footbed
- boxing and labelling of the box.

At the end of the above steps, boxes carrying an indication of the final consumer and the shipment destination are ready for their departure from the manufacturing plant.

The EUROShoE project was a huge technological endeavour to which many partners (shoe companies, technology providers, research institutions) contributed their know-how and dedication; more than two years have passed after its conclusion and some of the technologies that were explored during the project are progressively being industrialized and will soon reach full maturity.

Among the various merits of the initiative that can justify the huge economic investments that were made in it, we can say that EUROShoE contributed to:

- spreading the knowledge of the Mass Customization paradigm and of its application and benefits for the footwear sector,
- developing prototypes and demonstrators of the various enabling technologies that are needed to actually implement the paradigm, showing shoe companies that the technical challenges MC poses can be managed,
- demonstrating how the design, sale and manufacturing of mass customized shoes can be handled, showing that it can be done within the capabilities of modern technologies.

We believe the footwear sector is now becoming fully aware of this potential and that it is more and more ready to take advantage of this legacy.

3.3 Lessons Learnt and the EUROShoE Legacy

The EUROShoE project was the first "IP-like" project for the European Commission. In the sixth Framework Program (FWP), the so-called New Instruments were created called IP or Integrated Projects and the NOE or Network of Excellence. The IP projects were intended in particular to move a critical mass of R&D resources to really bring about a transformation in a specific industrial sector and as such were to be lead by industry.

EUROShoE was one project approved by one of the last calls of the fifth Framework Program (FWP) but it was designed to look like an IP of the sixth FWP in order to test the validity of the IP concept.

An IP needs to have a number of partners sufficient to bring a transformation in a sector and as such is normally made up of a large consortium; it should be led by industry and have clear focus on the end users.

We can say that most of these conditions were fulfilled in EUROShoE even if the framework conditions were not a true IP of the sixth FWP.

The consortium was large (34 partners) with six footwear manufacturers representing a good profile of the end users and of the European footwear industry. Many SME (both from end users as well as technology performers) were part of the consortium. We can say that the EUROShoE project was almost unique in the sense that:

- It had a vision - to develop, build and test the whole extended enterprise for mass customization of shoes (traditional shoes not the technical ones).
- It put the customer at the centre of the whole process - from the beginning with the 3D measurement of the feet up to the individual delivery of the shoes to the customer.
- It organized and set up a whole factory based on all the new results obtained in the project that included the point of sale, the design department, the last making, the sole making, the upper cutting, assembly and stitching, the final shoe assembly and finishing, the packaging and the delivery.

The project as an applied research one was certainly a success. Several new technologies were developed, implemented and tested. Other technologies were improved and tested but the main result was the total and almost seamless integration of all these technologies from the scanning of the feet to the digital data transferred from one department to another all the way down to the factory and the single machines and robots.

The core results of the EUROShoE project wre set in an old rubber boot factory in the centre of Vigevano, establishing therefore a ideal link with the glorious past of the capital of shoe making. The project has attracted a lot of attention at not only European but also worldwide level. It had also a role to try and explain not only to researchers but also to the more common people what are the benefits of a personalized and custom fit shoe. Very simple marketing material was used as shown in flyer reproduced in Fig. 3.7.

Certainly the television coverage at the final presentation of the EUROShoE project also helped to disseminate the idea that was mainly based on a new industrial paradigm for the footwear sector: personalized shoes that the common consumer (representing 70 to 80% of the global population) may buy.

The main innovations that form the EUROShoE legacy are:

- *"product"* - high tech shoes with personalized features
- *"processes"*- enabling technologies for all the phases of the product life cycle
- *"enterprise organisation"* - IT technologies, applications and new management tools (e-design, e-business, e-commerce).

The EUROShoE also had a social impact in employment, training and education:

- new jobs in the footwear sector and in related fields of shoe design, shoe machinery, management software and logistics
- new professional and higher skills;

and in quality of life:

- higher level of individual satisfaction (customized shoes at a reasonable price)
- role of the consumer - proactive versus passive
- health - better working conditions, fewer health problems and lower medical expenses
- environment - efficient use of direct (materials) and indirect (energy) resources.

More than two years after the end of the project, can we say that these goals and impacts have been reached? We can say that as in many research projects the results are not immediately applicable but some individual outcomes have resulted in patents and applications sold in the market (like a new kind of 3D scanner, software solutions for the fashion industry, flexible storage and transportation systems, etc.).The main idea of mass customized footwear has not yet taken its full place in the market but some good examples exist and many others are developing. The visionary idea that 30 to 40% of all shoes sold will be customized is still far away, but it seems that it will be like a snowball effect, once some big companies, in the trail of the few more courageous smaller companies, enter the MC market.

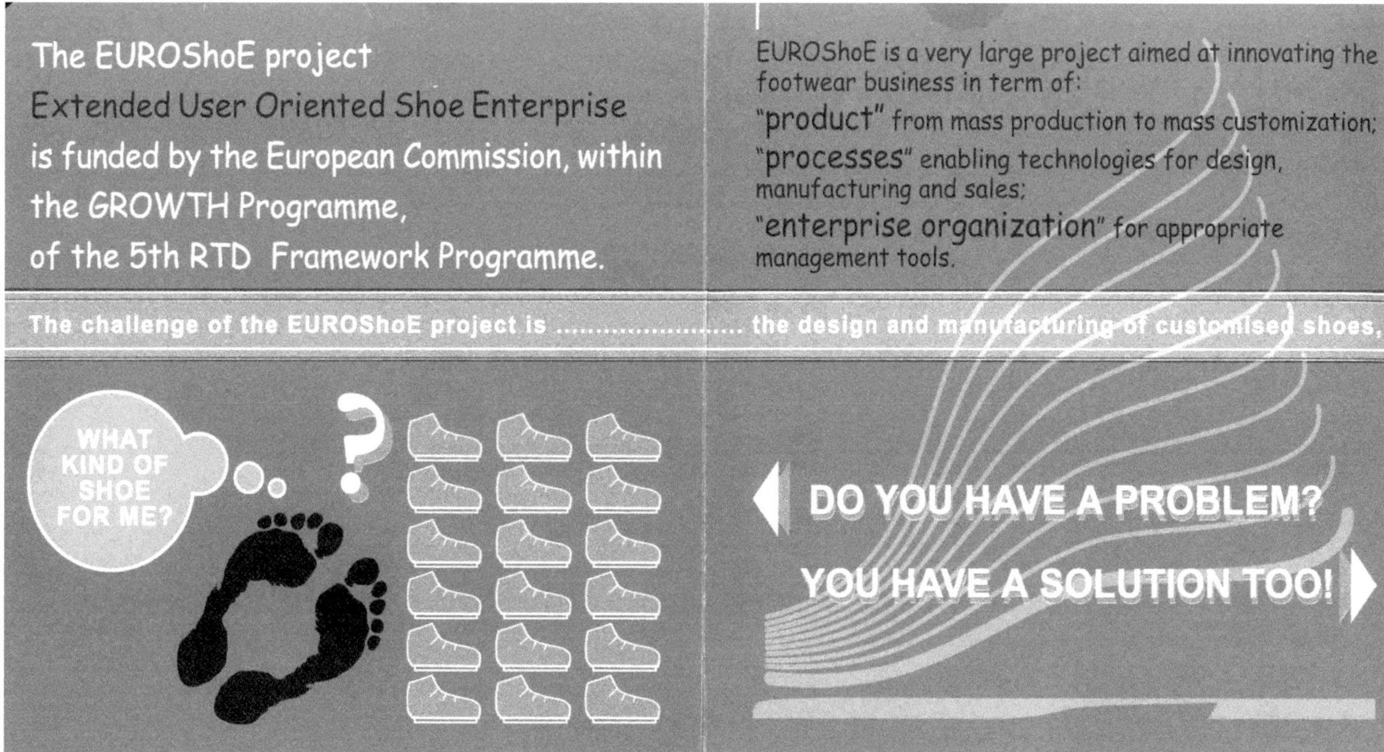

Figure 3.7. The flyer used to publicize the EUROShoE concept

4

Footwear Mass Customization in Practice

4.1 The Many Paths to Customized Shoes

There is no unique approach to footwear mass customization; the adoption of the new paradigm by interested companies must be "customized" case by case depending on their history, their current level of organization, and the priority the project has in their strategic plans .All these different "boundary conditions" have determined the various different aspects mass customization is taking in the footwear world.

So there are different "variants" of the model adopted by footwear companies; hence, prior to presenting the list of the all footwear companies that have already initiated their path in this new challenging domain and to presenting their specific case histories (which is the purpose of this chapter), we need to introduce some basic forms of classification. This is helpful to understand the peculiarities of these different experiences that we are going to present in the next pages and to highlight the advantages and drawbacks of the different implementations.

One very first classification (which we have already introduced in Chapter 2) is by *level of customization*: consumers' desires can be satisfied in different ways and aimed at different levels of consumer satisfaction. This level of satisfaction goes hand in hand with how much and how far (along which customization vector) the shoe is personalized to the specifications of the consumer. Since increasing levels of customization imply increasing levels of process complexity (see Section 2.3 for an in depth presentation), companies have elected to "enter" the mass customization world at one of these well-defined levels. They have chosen the approach that was most appropriate to their organizational configuration, to their brand positioning and to their tradition. This will be the first classification scheme to be treated.

Another aspect worth being analyzed is *the "origin" of the company* or, if we want, its past history; so a second classification criteria looks at where the Mass Customizing company is coming from: whether it belongs to the footwear world, or if it is a totally new start-up company, or if it is the result of diversification strategies in big luxury conglomerate or consumer goods multinationals. In addition the nature (type of shoes) and quality level of the product which is customized matter; so a third classification looks at the *product positioning* and highlights the implications of such positioning on the specific implementation model adopted by the company. Finally, a final classification can be made in terms of *operational approach.* This is a more complicated definition to be given; it deals with the technical modality the company adopted to handle the different processes that enable the sale, design and manufacturing of the customized shoes. In the next pages we will try to give a more comprehensive definition of these aspects and we will give a final classification of the shoe mass customizers in these terms too. We must remember that all these classification parameters are not mutually exclusive, but they rather represent the "attributes" of the specific implementation model that each company has, on its own, conceived to go along its personal (customized) path to mass customization.

4.1.1 Classification by Level of Customization

The choice of the level of customization is the result of a complex balance between how much and how far the company wants to satisfy the expectations of the individual consumer and the growing process complexity that this implies (for a detailed analysis of the relationships between level of customization and process complexity see Section 2.3). In Fig. 2.4, we have already introduced the notion of the "vectors" of customization, and we have explained that consumers' demands can be fulfilled along three specific directions: aesthetic, comfort and function. The four typical levels of customization (or at least the levels for which a definite affect on process complexity can be devised) deal with such vectors:

- *Style customization* - looks at satisfying consumers demands in terms of aesthetics, making sure that a shoe in the desired materials, colours and with the desired components can always be made for the consumer[28].
- *Best-matched fit* - aims at the privileging comfort (that comes mostly from the selection of the right shape of last) over aesthetics or, more appropriately, together with aesthetics; this first level approach to satisfying comfort demands, uses a simplification in terms of process (choosing a best-matching, already available last rather than manufacturing a tailor made one) to keep process complexity at a reasonable level.

---

[28] We must be careful with this definition: we do not want to mean that consumers have an infinite choice; they are rather guaranteed that their specific choice is always available within the range of combinations proposed by the producer (the so-called "solution space" see Piller [17]).

- *Custom fit* - is the ultimate comfort experience with the entire shoe (the last and all the components of the shoe) tailored on the morphometric data of the consumer.
- *Functional customization* - finally moves along the third customization vector that is the one of adapting the functional performance (typically for sport shoes, but not only) to the characteristics and demands of the individual consumer.

One aspect that we want here to investigate more in depth is the one of the relationships between the level of customization and the perceived level of satisfaction of the consumer. There is no exact or truly scientific definition that can be given of something like the "level of satisfaction"; these is possibly more easily estimated ex post by looking at how many of the clients of the company are and remain satisfied by the products they have bought and they come back again and again to buy new ones. On a very intuitive, yet possibly oversimplified, level we could define the "perceived level of satisfaction" of the consumer by stating that it relates to *how much a consumer feels to be pleased and attracted by the aesthetics of the shoes he or she buys, how much he/she feels comfortable wearing them when they are purchased and, afterwards, in time and, if this applies, how much his/her dynamic performance is eased or enhanced by wearing the shoes*

A second notion that is worth introducing here, once more based on the observation of what has happened so far in the footwear MC scenario, is that levels of customization are "additive". This means the higher levels include the lower ones; for example, aesthetic customization does appear alone, but if best-matched fit is offered, it normally includes aesthetic customization, as well as functional customization implies fit matching and aesthetics. With this in mind we can try to correlate the "perceived level of satisfaction" with the "level of customization" in the way illustrated in the Fig. 4.1. As it can be seen, the different levels of customization are positioned along the arrow at growing levels of consumer satisfaction:

- **Mass production** - any colour provided it is "black" (Fordist paradigm); left and right shoes are identical although left and right foot are not the same; size is chosen "experimentally" by the consumer, who must know his or her size, determining it with a "trial and error" procedure.
- **Style customization / shoe made to order** - the consumer receives a shoe that fits perfectly his or her aesthetic preferences (configuration process) and is guaranteed that the shoes will be available (since they are made to order) in the size specified. There is no risk that the shoes that are chosen from the aesthetic point of view are not available in the desired size; yet the consumers still has to know his or her size.
- **Best-matched fit / foot measured** - this level implies aesthetic customization; foot measurement (using a manual approach, or simple measuring devices or complex foot scanners) appears for the first time. The consumer is not supposed to know what size hi or her feet are, but identifying the right size and choosing the most appropriate last is the task of the producer (or of the shoe reseller). To make the process simpler the last is not really tailor made, but selected (matched) from a library of sizes as the closest one to the dimensions of the consumer; shoes are made to

order with the selected aesthetic configuration and in the size calculated after the measurement of the feet. This step produces a remarkable increase in the level of individual satisfaction of the consumer.

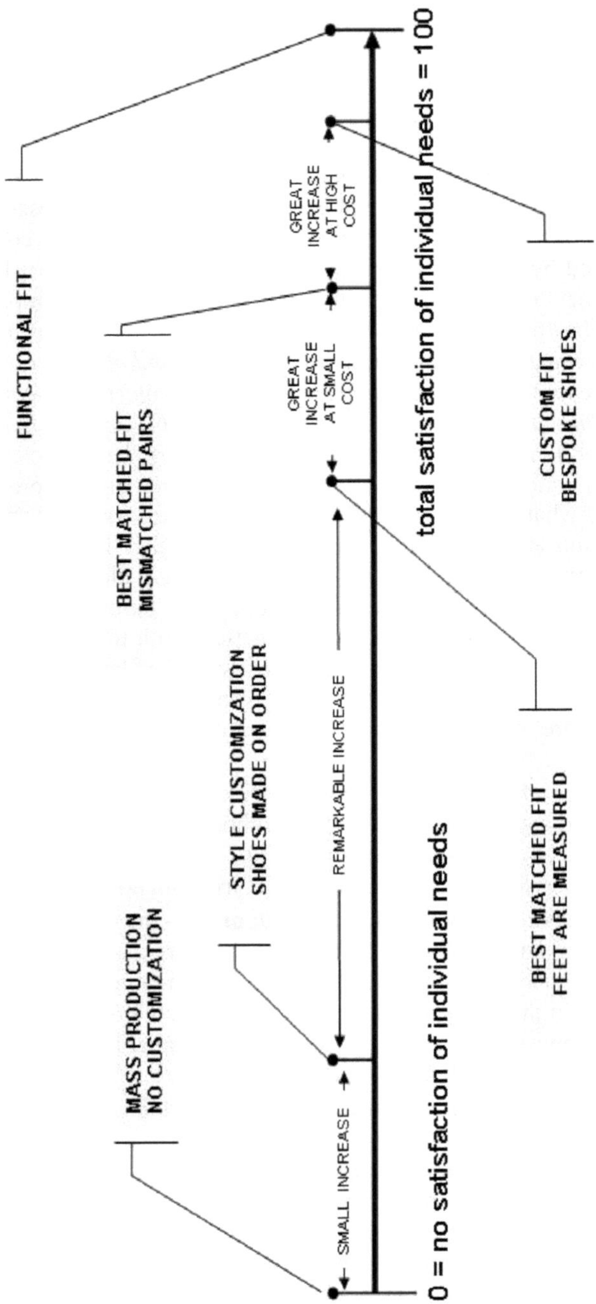

Figure 4.1. The level of satisfaction scale

- **Mismatched pairs** - same level as before (best-matched fit and aesthetic customization) but in addition to that, left and right shoes are of different sizes (mismatched pairs), to comply with the fact that left and right feet are different. A very pronounced effect in terms of the level of consumer satisfaction at a marginal additional cost in terms of process complexity (the producer must be able to handle batches of one single shoe).
- **Custom fit / bespoke shoes** - truly tailor made (bespoke) shoes in which an individual last is elaborated specifically on the foot morphology of the individual consumer; hence all the shoe components are also tailor made. At the border line of the mass customization paradigm, typically adopted by craftsmen. It also implies accurate foot measurements and aesthetic customization. Shoes are of course made to order with very long waiting time.
- **Functional fit** - a customization of the function of the shoe is added on top of size (best-matched fit) and style (aesthetic) customization. It represents the most complete form of footwear customization and it achieves, from the consumer's perspective, the highest level of individual satisfaction.

We will see in the following chapters which specific model (level of customization) the various companies that have launched projects in the field of mass customization have adopted to qualify their offer.

4.1.2 Classification by Origin of the Company

Another way of classifying footwear mass customizers is by looking at their origins and at the background of their companies. We have already introduced these concepts in Section 2.4; we will use here the same ideas to differentiate companies adopting the mass customization paradigm in shoemaking:

- *Innovative shoe makers* - firms already involved in the footwear business, with production experience, normally with their own brand, that believe in the market potential of mass customization and that are progressively moving towards the adoption of the paradigm (in one of the different forms described above) for their products. Here again there is no unique story: some companies have strong brands and are "flavouring" part of their normal product offer with the add on of customization; others have created separate and independent brands for the new product offer. Some control the retail side of the business, others simply offer their retail network the possibility of selling the new customized products. In discussing the various cases we will highlight these features for each one of them. What is common to all of them is the fact that they can all rely on available facilities especially for the manufacturing of the customized shoes.
- *Appealed outsiders* - these are typically start-ups, newly funded companies that were created by their founders specifically to exploit the potential of the new paradigm; there are cases in which the entrepreneurs behind the new venture had some previous shoemaking (or shoe design) experience. But there are also cases in which the founders had no specific

knowledge of the footwear business, but only a strong interest for the matter and precise business plans on the revenues that this way of approaching shoe consumers can generate.

4.1.3 Classification by Product Positioning

Shoe (mass) customizers can also be classified in terms of their product positioning or, more precisely, in terms of the final shoe price for the consumers they are targeting. From this point of view it must be remembered that "customized" shoes have always existed, all through the era of the industrial production (mass production), as highly expensive, exclusive products for a niche of elite clients (clients who can afford to spend in excess of € 1000 for a pair of shoes). The end price has a strict relationship with the size of the market the company is targeting at and with a true implementation of the mass customization paradigm. With reference to Fig. 4.2, we can distinguish between:

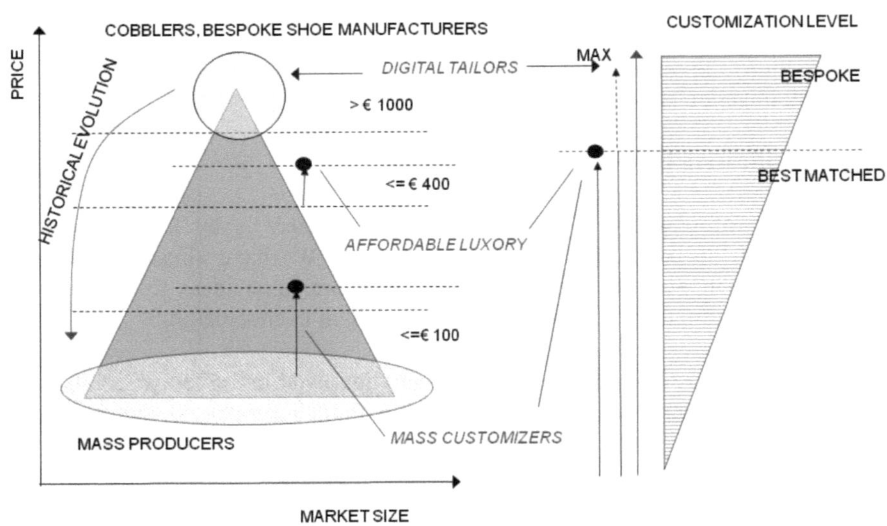

Figure 4.2. Product positioning in terms of price

- **True mass customizers** - these are the companies that are (or will be) mainly targeting at the mass markets where the highest potential, in terms of volumes, exists. They can be established shoe makers or new ventures offering their shoes in a price bracket that is such that their customized shoes can be bought by consumers with normal purchasing power, ready to pay a moderate extra cost with respect to the shoes they normally buy. There are offers in this segment both for sneakers, sports shoes and formal shoes. A typical street price level could be assumed to be around €100.00 per pair; better, the typical accepted price premium (the extra cost to pay for the customized shoes) is assumed to be no more than 10-15 % greater than the price of similar non customized shoes[29].

These products is positioned at the lower end of the market-price pyramid, and the level of customization companies like these are likely to offer, is up to best-matched fit (the majority of the current market players in this segment are in reality only offering aesthetic and style customization).

- **Digital tailors** - this second group of customized shoe producers is, viceversa, positioned at the vertex of the market–price pyramid. It is even questionable whether to call them *mass* customizers since the market segment they target, is the top quality–top price one: wealthy clients that can afford to spend more than € 1,000.00 per pair of shoes. Nothing new under the sun, in this respect; these are the same small workshops, craftsmen and cobblers that have been making truly bespoke shoes all along the last century, when shoe making, in general, evolved from craft manufacturing to industry. What do they have in common with the pioneers of mass customization? The interest for digital technologies and the adoption of modern equipment that can make their business more efficient and that can increase the level of service to their customers (for example in terms of delivery time of the shoes). We may invent a neologism, calling them "*digital tailors*", they still create their products with a predominantly manual approach, tailoring the shoes to the feet of their customers, but they make use of digital technologies to renew their old processes and to rejuvenate their business.

If those two "classes" of shoe customizers can be considered as evolutions of existing categories (mass producers of standard shoes and manufactures of bespoke footwear), the last group is a totally new breed:

- **Affordable luxury**[30] - their aim is different than both the one of the mass producers and of the digital tailors. Their effort is to make the luxury, normally offered by bespoke shoe makers, affordable if not exactly to the masses of consumers, certainly to a much wider segment than the one that can afford bespoke shoes. They position their products at a price level slightly higher than mass customizers (an average street price computed from the offers of those companies that are active in this segment is in the range of € 200.00 to € 400.00 with the latter as the top figure) but still far below the level of the truly bespoke shoes.

  They use modern technologies (also for manufacturing), in order to be able to offer high levels of customization (no less than best-matched fit) at

[29] For a detailed presentation and justification of these parameters see the survey done within the EUROShoE project [5].

[30] This definition was actually been first used by the German company Selve – whose case we will be presented in details in the following pages – to qualify their market positioning; we have adopted the term in this book to identify the group of companies that, in our opinion, share the same vision as Selve and have the same market positioning.

reasonable costs, yet investing in the quality of their shoes, the richness of materials and the fashion content in order to justify their top segment positioning. Their target customers are on the one hand the wealthiest among the purchasers of mass shoes, who cannot afford truly bespoke ones, but that are ready to pay something less than that for a high quality best-matched one.

On the other hand they can also "win" customers from the higher segments, being able to offer almost the same quality at a lower price (which is possibly not the most relevant driver) but with a better service: faster and more reliable deliveries, more configuration options, enhanced customer care. A potentially very promising new market segment.

In the analysis that follows, the companies that will be examined will be classified also from this point of view.

4.1.4 Classification by Operational Approach

One last possible classification of the shoe mass customizers currently present in the market is based on the technological approach they have adopted, as depicted in Fig. 4.3. We can classify the operational approach chosen by the company for its mass customization project using two parameters, calling the first one "*level of physicality*" by which we mean how much "physical" (for instance real shops versus Web sales) is the approach adopted by the company to structure the relevant processes of its business.

The second variable that is used in this classification is called "*level of control on MC process*". As we have described earlier in the book, MC of shoes relies on a certain number of processes, and we have also learnt that most of the traditional processes of the company have to be re-engineered in migrating to mass customization. Not all these processes are necessarily controlled directly by the company, which, on the contrary, may decide to manage some of them personally and to outsource others that are not considered to be either critical or strategic. From this standpoint, companies which have already started their journey in mass customization have made different choices.

Thinking of the two mentioned parameters, we can then identify four (nominal) categories (or groups of shoe mass customizers).

Pure Click Sellers of Customized Shoes

In this first quadrant of the graph we have companies that have chosen to concentrate on the sales process only and that have adopted, primarily, a Web-based (pure click) approach for their sales. If we look at the names and profiles of the companies in this group (each one of them will be described in detail in the following pages) we can see that we have both renowned global brands as well as newcomers, with local or continental dimensions. All of them are more interested in controlling what they consider to be the more strategic sales part of the business, which, in other terms, means the contact with the consumers. The Web approach is preferred (meaning a virtual shop with no real shops where consumers can go) although some of them (like Puma for instance) are using a combination of virtual and real sale points. In other words, they have gone for a combined strategy, in which Web sales seem to be predominant and sales in real shops are just a complementary channel[31].

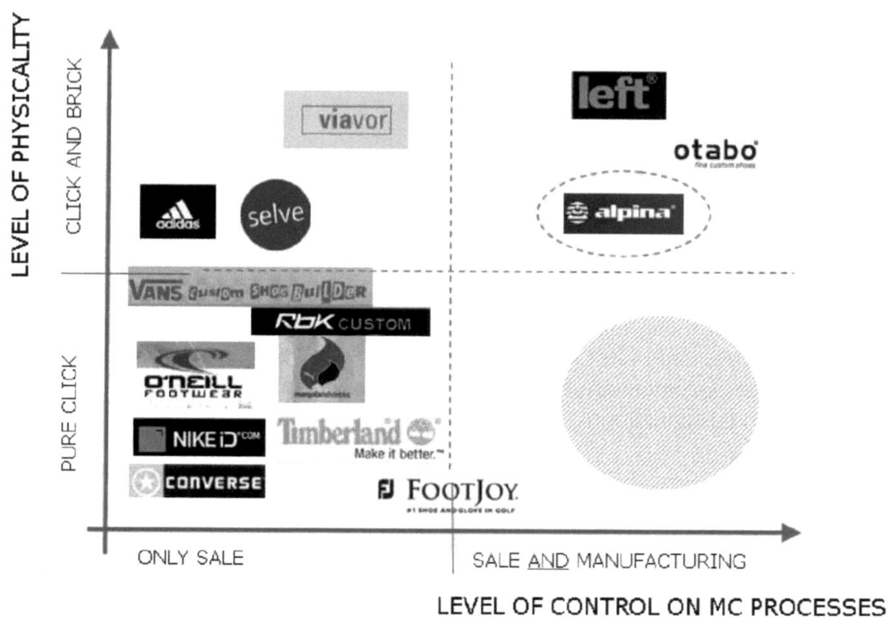

Figure 4.3. Classification by technological approach

Click and Brick Sellers of Customized Shoes

This second group includes companies that are again mostly concentrating on the sales part of the business, and that rely on manufacturing suppliers for the production of the customized (and for the big ones, also not customized) shoes. But, with respect to the names of the first group, their approach to the customers is physical more than virtual. They have started first with shops and put more emphasis on the direct personal contact with the consumers, also for the level of customization they have decided to adopt (most of them offer best-matched fit shoes that imply some form of foot measurement); this entry into the mass customization market obliges them to have real sales points where their equipment and sales personnel are based. As in the case of the "pure click", the Web sale channel is not disregarded, but this is now the complementary one: first real sales and real shops and then the Website.

31 It must anyway be noted that when it happens that these companies offer their customized shoes in real shops, these can be multibrand shops or brand ones; in both cases the line of customized shoes is sold together with the normal product offering of the company. In other words these shops are not dedicated solely to customized shoes.

Click and Brick Sellers and Manufacturers of Customized Shoes

This last group adopts, on the sales side, the same strategy of the "click and brick" sellers (so mostly sales in real shops complemented by sales via the Internet) but, contrary to the two previous ones, they have considered manufacturing to be a strategic asset and they control, in addition to sales, this process too. They own or control (through very binding partnership agreements) the production facilities that manufacture their shoes; in this way they can rely on precise delivery schedules and on flexible responses to their production demands, which are the prerequisites for a high level of service to their customers. We will see, in presenting the specific case histories of these companies, how they came to controlling or setting up such dedicated and optimized production facilities.

There is, in theory, a fourth group, to comply with the classification that we have introduced (the lower right quadrant of the Fig. 4.3); it should contain companies that adopt a "virtual" approach both for the sale and the manufacturing process. But since mass customization is about selling real shoes rather than virtual representations of them, the manufacturing process cannot be a purely Web-based one and real "nuts and bolts" facilities must exist somewhere. This is why there are no cases to be documented of a pure click, sale and manufacturing approach to footwear mass customization.

Figure 4.3 has already indicated the names of the footwear companies that are currently offering customized shoes to their clients, positioning them in the various quadrant of the graph depending of the last classification criterion which has been introduced. A brief account of their histories, of their specific MC implementation models and of their product positioning will be given in the next chapters for all of them; moreover, for a few exemplary cases, detailed case histories will be presented to document why and how these companies have moved to mass customization.

In dealing with the various cases, the level of customization will be used as primary classification criterion; nevertheless, in presenting them, indications will also be given regarding how these cases are classified following all the other criteria that we have described in this chapter. Hence the next sections are organized as follows: first companies only offering style customization (the most numerous group) will be presented. Then companies which have moved along the "comfort vector" of customization will be described, with specific case histories. Finally the case history of the only company (Adidas) that offers the highest level of customization (style and aesthetics, comfort and function; all three customization vectors together) will be treated in full detail.

4.2 The Champions of Mass Customized Shoes

We will concentrate in the next few chapters on those footwear companies that have, in the recent past, started projects in the domain of mass customization and that are currently offering their customers personalized products. The main purpose of this exercise is to provide traditional footwear companies or potential start-ups with guidelines, derived from the examination of what can be considered "best practices" in this field, on how to tackle this complex matter in a realistic and successful manner.

Although our focus will be on companies which have survived the initial pioneering phase and that are then progressing along the path of mass customization, we believe that also the few failures that were registered until today are worth a mention. Actually we believe they deserve more than a simple mention because it is true the sometimes there is more to be learnt in the reasons of a failure than in the examination of several success cases. Learning from mistakes others have made in the way they have approached the topic of shoe mass customization will give precious indications on errors to be avoided in order to make the mass customization project a success.

Although mass customization in itself is not a new concept (see [1] for definition of mass customization), the idea of applying it to footwear is certainly relatively more recent. In the early nineties the first concepts (and the first enabling technologies) were proposed; they were conveying the idea that shoes could be personalized and customized for each individual consumer, but with an approach that could anyway preserve the economies of scale typical of mass production and that could take advantage of the technological advancements that those years were overseeing. The growing interest for the topic is also testified by the number of patents that were filed, both in the USA and in other countries, to protect the "idea" (the concept, the process, the approach to the consumers) of mass customized shoes and of personalization of footwear.

The first industrial attempts date back to 1993-1994 when an American company, named very appropriately *CustomFoot*, first appeared on the market offering customized men shoes to their clients in the USA and having the shoes manufactured in Europe (presumably in Italy). The company did not live long and soon disappeared, but it can be registered as the first practical attempt to conjugate footwear customization with affordable costs, addressing its offer to masses of consumers rather than to the usual elites.

Another small group of forerunners followed in the last decade of the twentieth century with different projects, mainly aimed at exploring and assessing the potential of the apparently promising new market. Some of thse companies failed (and a brief account of the most illustrious victims will be given later), but others survived, as was the case of NikeID, the customization brand of Nike, which we will discuss in some detail in the next section.

It can be noted that the majority of the companies that approached the new paradigm in the last decade of the past century (1990-2000) did so by taking full advantage of the new emerging Internet world that was developing in that decade to become the social phenomenon that we all know today. So it is no surprise that such experiments were mainly focused on offering style customization through Web sales only (or primarily). When some of the big names in the sneaker field moved on along that way, many others followed; some of them are still there.

A key enabling technology was, on the other hand, needed to move from Web sales only and style customization to the next level of customization (fit and comfort): foot scanning. The first electronic devices to measure the foot and to capture its relevant dimensions (the so-called foot scanners) appear in the second half of the decade 1990-2000[32]. Their availability favoured the start up of new ventures, which were offering comfort (fit) customization together with style personalization, showing consumers that higher levels of personal satisfaction were achievable and could be offered.

The origins of the current market leaders (the mass customization champions) date back to that period; many of these companies started exploring the new scanning technologies towards the end of the century (2000) and had begun elaborating their mass customization strategies around that time. Most of them are still active today, having further expanded their operations, improved their business models and developed, in some cases, a whole range of specific enabling technologies that they considered strategic for MC. Others were lost along the way; for many of them their presence in the market was too short or sporadic, so it is impossible to give a full account on all the births and deaths that were registered in the last 15 years.

If we look at the market situation as it is now (by "market" we mean that of customized shoes), we can see that all levels of customization as introduced in the previous chapters are present: the "champions of footwear mass customization" have different origins, offer products that address different market segments and have adopted "personalized" approaches for the implementation of the paradigm. We will illustrate how they came to that and how they are performing now. We will see how big brands, small start-ups and newcomers are active today. We will also see that their number is still very small compared to the many thousands of shoe companies that exist in the world; but it is worth noting how this number has been growing in the recent years, to indicate a clear trend of an increasing interest from companies for the mass customization model, very likely sustained by a parallel increasing interest and motivation from consumers. We will further comment on that in the final sections of this chapter.

4.2.1 The Vanished Pioneers

As stated earlier, failures in this field always teach lessons to be learnt; there have been many of these cases, even in the relatively short history of footwear mass customization and presenting all of them in detail would not be possible. But at least two cases are worth being treated in some detail:

1. **Creo**. The German company Creo Shoes was one of the first to try exploiting the new paradigm. Creo started its operations at the verge of the new millennium, choosing to use the Internet as the prime and unique sales channel (no shops and no investments in fixed assets of any kind). They offered casual shoes, similar to sneakers but with some interesting

[32] In reality, electronic foot scanners wouldn't be strictly necessary to measure the foot and a great variety of manual systems or mechanical devices already existed to perform the measuring task and to identify the size of the foot (like the well-known Brannock device); but certainly automatic scanners, which required no skill and were faster in measuring the foot, did represent the enabling technology that triggered a new round of enthusiasm for footwear mass customization.

features like a strap closing and original colour and material combinations. Shoes could be ordered only at the Website, which featured one of the first "shoe configurators" to be seen in the market. The company had no proprietary manufacturing facilities, relying entirely for the production side of the business on subcontractors (located, based on the information that it was possible to gather, in the former Eastern bloc countries). Creo did manage to launch its products and gain a relatively large consumer base; certainly they had the merit of "educating" consumers on, at that time, new ideas like ordering shoes via the Internet, personalizing the aesthetics of the shoes by playing with colours and materials (style customization). In 2001, after a few seasons of sales, the company stopped its operations and went out of business. An *ex post* in depth analysis of this case offers many explanations for this failure, among which the quality of the manufactured shoes (more from the side of the pure construction quality than on that of the design, which was, all things considered, original and pleasant) did count. Creo experienced many of the weaknesses and difficulties of the new business approach, like the fact that consumers had to place "blind" orders and could only asses the value of their purchase when the shoes were delivered. This is a very critical issue that strictly relates to the quality of the products; even minor mistakes or failures are not forgiven and they immediately transform into lost customers. Creo had to develop strategies for that and had virtually no time to fine-tune them. But more than that it is our opinion that one of the main reasons for not being able to reach a breakeven in terms of sales, was possibly to be found in the low number of shoe models which were offered (basically one) and the relatively low quantity of configurations which were offered; all that was not enough to stimulate old customers to buy more and to gain a high enough number of new clients to sustain the operations of the company. A lessons learnt for posterity.

2. **Customatix**. We have decided to present this second case because the reasons for this second fiasco are opposite to the ones of the previous story. If Creo failed because of the limited choice (and quality) of their customized shoes, Customatix failed possibly because of too much choice. We have no precise date about when the company appeared, offering shoes that could be bought from their Website (so no real shops in this case too) and configured through a very complex and somehow confusing product configurator, but Customatix did manage to live some years longer than Creo (their Website was still open until at least 2004). As in the case of the German pioneer, Customatix did not have its own factories; all of the shoes that the company sold were produced in China and shipped from there to the various clients in the world, directly to their private address. Customatix offered sports shoes and sneakers, in literally hundreds of different models and, by combining models with colours and materials, in billions of possible combinations (as the company very proudly claimed on the home page of its Website).

> This turned out to be an overwhelming choice for the consumers, who run the risk of getting lost in this multitude of different products and who badly needed some sort of assistance from the producers to guide them through the various choices (assistance that was indeed introduced in the form of a sort of purchase advisor, but not at a level that really overcame the difficulties the consumers were facing). The company existed for a few years, detailed data on their sales volumes are not available, but as a serious analysis on their profitability and on the size of their business is not possible; looking for them in the Web today would give no answer. Customatix seems to have vanished (but some traces of their adventure can still be found in blogs and in a few scattered Web pages). The main reason for this second negative example is, in the opinion of the authors, having offered consumers too much for too little difference with respect to other competitors in the same family of footwear, with no real distinctive values in their line of products. It was assumed that giving consumers an almost infinite choice to configure and to design their shoes would have been, in itself, the winning factor and would have induced thousands of consumer to chose Customatix and to buy their shoes, but this exaggerated level of choice eventually proved to be a mostly negative aspect of the business model of the company.

What these two cases tell us is that "mass customization", *i.e.* the possibility for the consumers of building their own products (at least form the aesthetic point of view), does not pay by itself, but it is not a good enough reason for clients to turn to the new business. Also when the price is not an issue (both Creo and Customatix were offering their shoes at a fairly reasonable price compared with similar products in the same category), MC must offer more; the added value for the consumer of the new approach to buying shoes must be more clearly evident and a very careful balance between mass customization (appealing range of products, with and appropriate number of combinations; manageable both for the producer and the consumer) and mass "confusion" (a too large choice with enormous handling costs for the producer) must be found.

If failures have lessons to be taught, success stories are certainly more attractive and they can show which paths to mass customization are to be taken in order to fully exploit the new paradigm. The companies that we are going to present in the next chapter all shared the same vision of the market potential for customized goods (in this specific context we refer to shoes, but some of them that are also active in producing other goods, are considering similar projects for these goods too), they have started the first explorations at the beginning of the new millennium (around year 2000), and they are now pursuing this strategy in a very determined way.

From a time perspective, we can notice two groups: the forerunners, which started their project back in 2000 and are today in a second more mature phase of their MC approach, and the followers. The second group is made of companies that began with mass customization more recently, on the wave of the growing interest showed by consumers and on the success of the forerunners. It is worth highlighting how most of the companies of this first group have two very evident factors in common: they only offer style customization (with the exception of Adidas) and they are all involved in the manufacturing of sport shoes, casual shoes and sneakers. A short list of these companies is presented in Table 4.1.

Table 4.1. Producers of mass customized sports and casual shoes

Company	MC project name	Website
Reebok	Reebok Custom	www.rbkcustom.com
Converse	Converse one	www.converse.com/conversone/
Nike	Nike iD	www.nikeid.com
Puma	Mongolian BBQ	www.puma.com/mongolianbbq/
Timberland	Customboots	www.timberland/customboots
JG Customs	JG Customs	www.booktwon.com/jgcustoms
Vans	Vans	www.shop.vans.com
O'Neill	designyoursneaker	www.oneill-action.com
FootJoy	Myjoys	www.myjoys.com
Adidas	Miadidas	www.miadidas.com
Steve Madden	Design your own...	www.stevemadden.com

All the companies mentioned in Table 4.1 offer their customers several lines of customized shoes; they do that with different modalities and with different styles, but they all seem determined to pursue in their customization strategy. Others are expected to follow, and there are indications of further firms active in the same sector that are seriously considering the possibility of entering the MC arena. We will give some accounts of these future projects in the later sections. Sports shoe manufacturers are not the only ones to be attracted by the idea of mass customization; other producers are offering dress and formal shoes too, with a further specialization between men's and women's shoes. As in the case of the sports shoe manufacturers, there is a great degree of homogeneity in the approach to mass customization that all these companies have chosen: all of them offer comfort (fit) customization together with style personalization and they seem to privilege serving their customers in real shops rather than addressing them to their website only. A short list of these companies is presented in Table 4.2.

The aim of Section 4.3 is to present in details the case histories of each one of the 16 companies that appear in Tables 4.1 and 4.2 so that these examples of "best practice" can furnish the basic guidelines for a successful implementation of the mass customization paradigm.

It was not possible to gather the same level of detailed information for all of the companies in the lists, so we have decided to just give a brief description of the main aspects of the MC projects for the majority and treat in detail only the most significant ones.

Table 4.2. Producers of mass customized dress and casual shoes

Company	Type of shoes	Website
Otabo	Men's	www.otabo.com
Selve	Women's	www.selveshoes.de
Leftfoot	Men's	www.leftfootcompany.com
Viavor	Men's (and women's soon)	www.viavor.com
Alpina	Trekking boots	www.alpina.com

A complete analysis of these cases would have required also an examination of some economical indicators that measure the profitability of the new business (like volumes of yearly sales, costs versus revenues of the specific MC business unit or of the company as such, ROI data and similar). Collecting these sort of data has not been easy, not to say impossible. They are considered by the companies as "sensitive data" so that virtually none of them was prepared to share with us this information Hence the only possible "success indicator" that we can use in assessing the outcomes of these mass customization projects is their longevity and the fact that the companies still actively carry on with their original strategies.

4.3 Case Histories

In presenting the case histories of the champions of footwear mass customization (according to the considerations that we have made in Section 4.2), we had to chose from among the various possible classification criteria (see for their definitions Section 4.1), the most immediate and intuitive way to analyze the different implementation models adopted by the companies. We have chosen the "level of customization" (see again Section 4.1).

Hence, the next three sections will present the case histories of the 15 footwear mass customization champions according to the following subdivision:

1. *style customizers* (vector of customization: aesthetics)
2. *fit customizers* (vector of customization: comfort), which also includes style
3. *function customizers* (vector of customization: function), which also includes fit and style

In describing their specific cases we will also give indications on the most relevant aspects of their MC projects, following the other classification parameters introduced previously: the company's origin, the technological approach adopted and, last but not least, its product positioning, both in terms of type of shoes offered (men's or women's, sport, casual or dress shoes) and average selling price.

This basic set of data will be provided for all the companies together with some brief notes on the history of the company itself and on the origin and peculiarities of the mass customization project; in addition to this general level of treatment, three detailed case studies are included in the next sections. These studies are dedicated to companies whose mass customization projects were considered to be of a particular relevance and hence deserved a more in depth presentation. These cases are:

- *Leftfoot*, as the typical case of a traditional footwear producer that has seen in mass customization the way to renovate its old fashioned and traditional business and to regain competitiveness
- *Selve,* as the only producer of customized women's dress shoes and as the typical example of a start-up company that was funded specifically with the aim of exploiting the business idea of mass customization.
- *Adidas*, as an example of a global and renowned brand that has decided to explore the MC paradigm at the highest level of complexity in accordance with its quality and technological tradition.

We will give for these three cases a full account, mostly based on recent data, specifically collected for this purpose with face to face interviews with the company owners or the managers responsible for the MC strategies and for the implementation of the project.

The presentation of these cases is structured along a predefined format that explores the many aspects of the implementation of the MC project, covering a list of topics of a great relevance. The aspects that are presented refer to the following:

- **Company profile** - the basic data of the company are presented, with information also on the typology and volumes of production, and basic data on its main facilities and its origin.
- **Passage to customized products** - this second section explores the motivations of the company in adopting the new paradigm: why it was considered a strategic step to be taken and when this step was undertaken.
- **Customized products** - this is the section dedicated to presenting the product offer of the company, the kinds of shoes it manufactures, the specific level of customization offered and the way the customization process is presented to the consumers.
- **Process for customization** - this part of the template present how, in details, the most relevant ones of the customization process were implemented, which transformations had to be made on the traditional processes of the company or how, when the company was a start-up, and how the new processes were planned and organized.
- **Market entrance of customized products** - this is the section in which the product positioning is analyzed. Information are also presented on how the company introduced the new product offering, how the message was and is transferred to the market, and how the market penetration is taking place.

- **Customers' perspective** - this part of the analysis is about how customer perceived (and do perceive now) the new product offer in terms of advantages (and disadvantages) for them and in terms of competitive factors for the company.
- **Case assessment** - the most complex (and delicate) part of the exercise is the one of giving a final assessment of the case; some basic considerations are made here (case by case) in terms of points of strength and weaknesses of the specific company case and also in terms of threats and opportunities (following the approach of a simple SWOT analysis).
- **Future perspectives** - a few comments on the future plans and perspectives of the mass customization operations of the company will be given; this also includes the final evaluation on the future potential (in the view of the company) of the mass customization business model.

As we have mentioned at the beginning, only three such cases are contained in the book (Leftfoot, Selve and Adidas); the remainider (12 cases) are treated at a slightly lower level of detail, yet still providing all the most relevant information about the companies and the particular mass customization approach they have adopted. We believe that the bulk of this material can provide both the occasional reader and the footwear expert willing to know more about the MC paradigm with a valuable insight on the current situation of mass customization in the shoe world.

4.3.1 Style Customization

We will cover in this section the cases of those companies that offer customized shoes for which only aesthetic personalization is possible. This is the positioning, in terms of level of customization, that groups the highest number of firms. Although there are differences in the way each one of them has actually implemented the paradigm and operates its new business, there are some basic aspects that are common to all of them:

1. The consumer is supposed to know (and the information must always be provided at some stage during the sale process) the size of shoes he or she wears. Hence there is no foot measurement foreseen in the process;
2. Shoes are then "configured" both at the Website of the company and, when the products are also sold in normal shops, using computers in the shops which run the same configuration program. The "configurator" (the software application that allows the consumer to create its own version of the shoes he or she wants to buy, by combining the different components, materials and colours the producer has made available) plays a key role in the entire process and is certainly the most important enabling technology to support it.
3. If the sales transaction is concluded by the consumer (which means that the shoes are actually paid before the manufacturing process starts) a production order is placed and the shoes will be manufactured and then shipped to the consumer's address (or to the shop when it exists).

All the cases that we are going to cover in detail in the next pages have the above steps of the sales process in common; as an additional commonality factor, they all belong, in terms of type of shoe they produce, to the same market segment (as we have already noted in the previous chapters).

The survey that follows presents an overview of all the active offers of customized shoes at that time (in this case style-aesthetic customization). Due to the growing interest in the matter and the constantly increasing number of new projects in this field, it is possible that not all the names of the shoe customizers actually active are referenced here.

The list that appears in the following pages must then be considered as indicative of the situation of this market at a given point in time, rather than an exhaustive and complete treatment of all the cases that can possibly be documented. Moreover, whilst for some of the cases information was available and a detailed coverage of the case was possible, for some others complete information was lacking and only a general overview was possible.

1. Reebok– Reebok Custom [date of survey: October 2006]

CUSTOMIZATION LEVEL	style–aesthetic; size declared by consumers
ORIGIN OF THE COMPANY	innovative shoe maker; global brand
PRODUCT TYPE	Sneakers; casual shoes
PRODUCT POSITIONING	true mass customizer
OPERATIONAL APPROACH	pure click; no shops; outsourced production

Reebok joined the group of the style customizers only recently; the dedicated Website exhibits one of the best configurators currently available with many features; it can be considered almost a perfect example of how a configurator should look (Fig.4.4). Nevertheless the risk exists that Reebok Custom is just seen as a late follower, copying what everyone else in their industry does. All in all there are no innovative customization features from the perspective of the consumer that can mark a difference with respect to its competitors.

The range of customized products includes four main models (just enough to make the offer attractive from the standpoint of the consumer), yet the four models look all very similar and the difference in the construction of the four shoes is not so evident. Once a model is selected and the gender and age of the consumer is chosen (male, female or child), the foot size is requested and the configuration process can start. The visitor of the Web shop has the choice to "start blank" or can be inspired by looking at some predefined configurations or select from some "hot picks" produced by a shoe designer. Four different portions of the shoes can be selected with a very high number of details that can be configured within each portion.

The main features of the components as well as the colours of the various portions can be chosen by the consumer, among a very long list of different colours.

Figure 4.4. The Reebok Custom Website and configurator

The materials used don't seem to be selectable; but even some minute details (like the width of the laces) can be defined. Designs can be saved and shared with other users.

At the end of the configuration process, at the checkout, the final price of the configured shoes is presented; the price range is such that Reebok Custom can be included among the true mass customizers, aiming at the mid price bracket (roughly between US$75 to US$125 per pair) typical for this kind of shoes. There doesn't seem to be, by comparison with similar non-customized products, any actual price premium for the customization service. The dedicated Website of the company (Reebok Custom) seems to be the only sales channel possible; we are not aware of sales of these shoes in Reebok's traditional sales outlets. As far as manufacturing is concerned, these shoes are presumably handled by the same facilities (very likely based in China or in general in the Far East) that produce all the other shoes sold by the company. A classical "pure click" example.

2. NIKE – Nike iD [date of survey: October 2006]

CUSTOMIZATION LEVEL	Style-aesthetic; size declared by consumers
ORIGIN OF THE COMPANY	innovative shoe maker; global brand
PRODUCT TYPE	high end sports shoes and sneakers
PRODUCT POSITIONING	true mass customizer
OPERATIONAL APPROACH	pure click; no shops; outsourced production

The NIKE iD Website (Fig.4.5) is more a sort of supermarket customization than only a place where customized shoes can be purchased: 51 shoes (31 for men, 17 for women and 3 for children) 6 bags, 5 watches and 3 kinds of balls, apparel and various kinds of gear can be found in the NIKE iD shop.

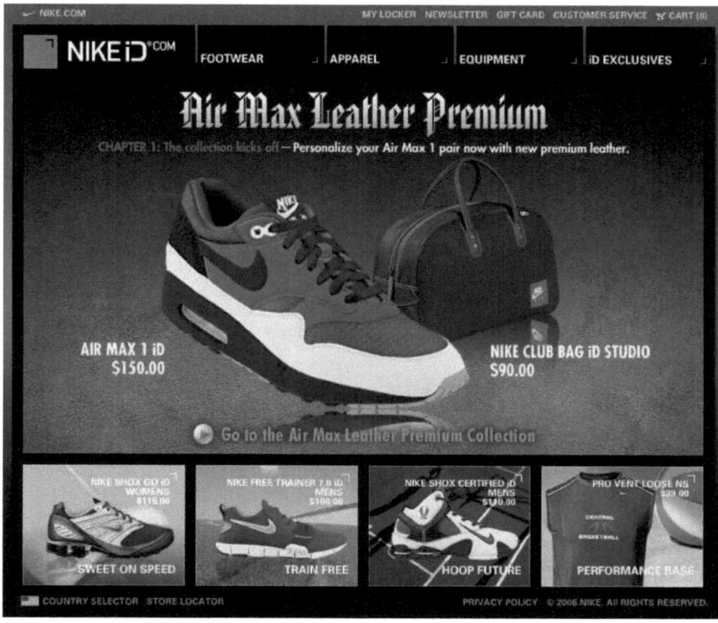

Figure 4.5. A section of the NIKE iD website

It looks as if the customized products sale portion is somehow "swallowing" the entire B2C operations of Nike. The third generation of the Nike iD Website also features a section on the "iD exclusive" in which the most fashionable metro collection can be seen and customized to the desires of the individual customer.

An overwhelmingly rich choice of shoe models (ranging from running shoes, to soccer boots, from casual shoes in general to fashionable urban styles) are available as part of a complete "customization frame" that includes accessories, apparel and gear of various kinds, this is much more than simply selling personalized footwear. This is a clear indication of the decision and commitment with which Nike is pursuing its product customization strategy, as a legacy of its position as forerunner in this field.

Once the shoe model is selected, the process looks pretty much the same as the Converse one (or perhaps *vice versa* - the Converse approach is derived from the Nike one considering how much earlier Nike started its explorations in the domain of mass customization); the configuration engine has the same look and feel, with the many different colour choices that are possible for the different parts of the upper and of the outsoles. Nike too gives the possibility of personalizing the shoes with a unique ID specified by the customers (different for the left and right shoes) that will be embroidered on the back part of the upper. The price positioning is higher than the previous competitors (as it is the case also for the non-customized products), ranging from approximately US$80 to something like US$120-US$150. As usual, after a firm purchase order is placed, shoes are manufactured in the normal production facilities of Nike in China and the Far East and then delivered directly to the consumer.

3. Converse – Converse one [date of survey: October 2006]

CUSTOMIZATION LEVEL	Style-aesthetic; size declared by consumers
ORIGIN OF THE COMPANY	innovative shoe maker; global brand
PRODUCT TYPE	low top-high top sports shoes
PRODUCT POSITIONING	true mass customizer
OPERATIONAL APPROACH	pure click; no shops; outsourced production

Figure 4.6. The Converse One Web page

Three shoe models (Chuck Taylor high and low top, Jack Purcell low top and the classic One Star) can be customized (Fig.4.6). Custom colour and materials and embroidered lettering online using the same approach as Nike iD are offered. The products offered are mostly basketball shoes and boots and casual shoes taken from the well-known Converse catalogue.

The range of choice is fairly good (three models, with seven variants in terms of top height, and four material options: canvas, leather and suede plus the classical Converse red canvas); it is presented as a first step at the beginning of the configuration process, giving the consumers plenty of choices right at this step. Prices for the various models (ranging from US$45 to US$88) and materials are immediately displayed so that the consumer has an early indication of how much the personalized shoes will cost.

The base choice for the material to be used is made at the beginning of the process. After having declared the foot size (given the declared shoe size, normally used by the consuner, the system indicates which Converse size should be ordered), the consumer is taken to the colour selection section of the configurator; all separate components of the upper (sides, tongue, lining, laces) and of the rubber sole (sidewalls) can be configured, choosing from a very rich palette. At the end of the process a personalized ID can be indicated and it will be embroidered either on the heel stripe or on the side of the upper (unique identification of the shoes).

The Converse approach resembles very closely the Nike one, also with many commonalities in the steps of the configuration process and the technical features of the configuration engine. As in the cases treated so far, the customization process is only possible via the Web and there are no shops in which it can be done in the "real" world. Once the purchase is confirmed by the consumer and the checkout process is concluded, a production order is placed and the consumer will receive his or her pair of shoes at the address specified. Manufacturing is taken care in the traditional way, relying on the manufacturing facilities of the subcontractors Converse normally works with.

4. Puma – Mongolian Shoe BBQ [date of survey: November 2006]

CUSTOMIZATION LEVEL	Style-aesthetic; size declared by consumers
ORIGIN OF THE COMPANY	innovative shoe maker; global brand
PRODUCT TYPE	high end sports shoes and sneakers
PRODUCT POSITIONING	true mass customizer
OPERATIONAL APPROACH	click, and a little brick; customized products via Web and in shops; outsourced production

Puma adopted a very peculiar approach to mass customization with its Mongolian Shoe BBQ product offer: a single style "served up" both via the Web and at hands-on kiosks open for limited times at selected Puma locations. A very tactile interaction with an interesting example of a "physical configurator" (Fig.4.7). Puma does the same things as all the others but with a different "flavour", creating an oriental atmosphere that characterizes the style of both the Website and the real shops.

The consumer is basically offered the same possibilities that the other forerunners of style customization do, but only one model is available for personalization; customers are driven through the configuration process, which gives them the possibility to select (once they have declared the size they want their shoes made in) from a very rich palette of colours and materials for all the 15 components (including outsole and midsole, upper and laces) the shoe is made of. But instead of doing it with the normal IT-like systems that all the others use, Puma uses the "metaphor" of the Chinese (Mongolian) restaurant – in this the parts to be chosen are served on a tray, with colour and materials picked from colourful and tasty bowls with the help of chopstick. When all the ingredients are chosen the consumer can pass his or her personal recipe to the "cook" to be prepared (manufactured). A very intriguing atmosphere that in fact hides a rather complicated and relatively time-consuming process.

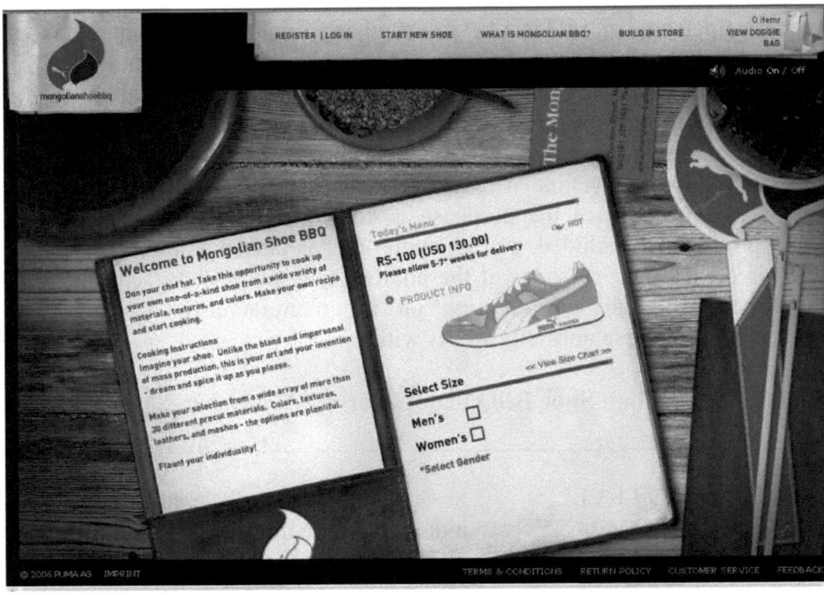

Figure 4.7. The opening "menu" of the mongolian shoe BBQ

with a manual configuration process, with real trays, real samples of the different parts of the shoe in different colours, and real "refrigerators" in which precooked (fully configured) shoes are exhibited as "inspirations" for the clients visiting the shop. The way Puma is exploring mass customization is certainly very intriguing and has a great appeal to consumers, who are stimulated to try playing around with the concept, perhaps more enjoying the process as such than the shoes they produce with it (an interview with the sale personnel in one of these shops confirmed that the actual rate of purchases at the end of the game is not that high, but that "*people do enjoy it very much*"). A purchase order is produced at the end of the configuration step; shoes are then manufactured in the same facilities Puma uses for its standard shoes (mainly located in the Far East).

All things considered, it looks as if Puma is still at a very initial step in its exploitation strategy of the mass customization paradigm; one single style is probably not enough to really motivate consumers and the configuration process takes too long for a "massive" approach to sales. At the current stage it looks more an experimental analysis of consumers' behaviour analysis than a true market alternative to mass produced footwear. The BBQ price is US$130 per pair, with a price positioning not very far from the standard Puma products (no relevant price premium). The delivery time is indicated to be five to seven weeks, which appears to be quite a long period with respect to the typical expectations of consumers interested in customized products.

5. Vans [date of survey: November 2006]

CUSTOMIZATION LEVEL	Style-aesthetic; size declared by consumers
ORIGIN OF THE COMPANY	innovative shoe and accessories retailer
PRODUCT TYPE	Casual shoes and sneakers
PRODUCT POSITIONING	true mass customizer
OPERATIONAL APPROACH	click and brick; web and real shops with outsourced production

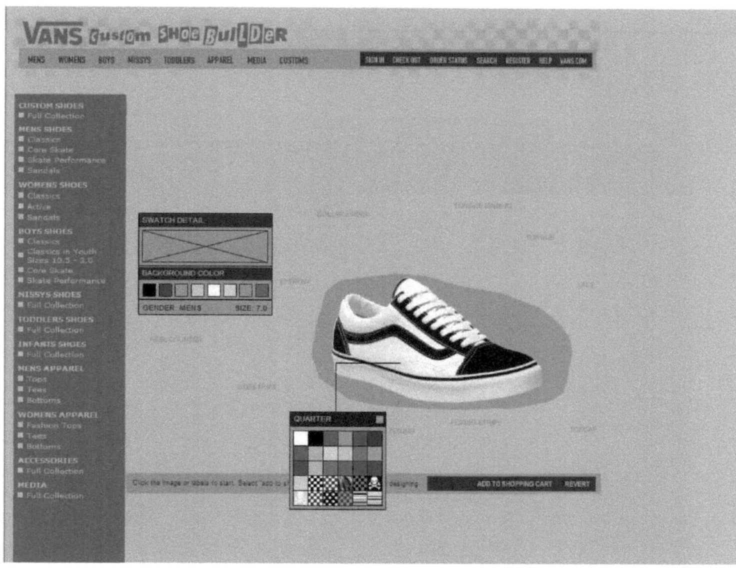

Figure 4.8. The custom section of the Vans Website

Two customizable models of shoes are offered. Custom colour and patterns online are possible, with a solid and well thought out interface for colour selection. Van is an American shoe retailer, offering various models of slip-on and snicker shoes for men and women, from infants to children and boys and girls, with a well-established and organized e-shop where shoes of various kinds can be selected from a huge catalogue and ordered. They have recently introduced a dedicated section of their online shop to "custom shoes"; two models are offered and full style configuration is possible, by selecting materials and colours for the various parts of the upper and the outsole.

A pretty neat and easy-to-use configurator (Fig.4.8) that only partially compensates the relatively low product offering of customized shoes. No information is available on where Vans shoes are manufactured and on how the retailer organizes the specific supply network for customized shoes; an easy guess would indicate the Far East and China as the most likely areas of shoes production.

The price tag for Vans custom is US$60 plus shipping costs (shoes are delivered to the address specified by the client) with a delivery time of three to five weeks; in this case too it is assumed that such a long time is acceptable for the customer.

6. **Timberland – Custom Boots** [date of survey: November 2006]

CUSTOMIZATION LEVEL	Style-aesthetic; size declared by consumers
ORIGIN OF THE COMPANY	innovative shoe maker; global brand
PRODUCT TYPE	high quality (and cost) boots
PRODUCT POSITIONING	true mass customizer
OPERATIONAL APPROACH	pure click; no shops; outsourced production

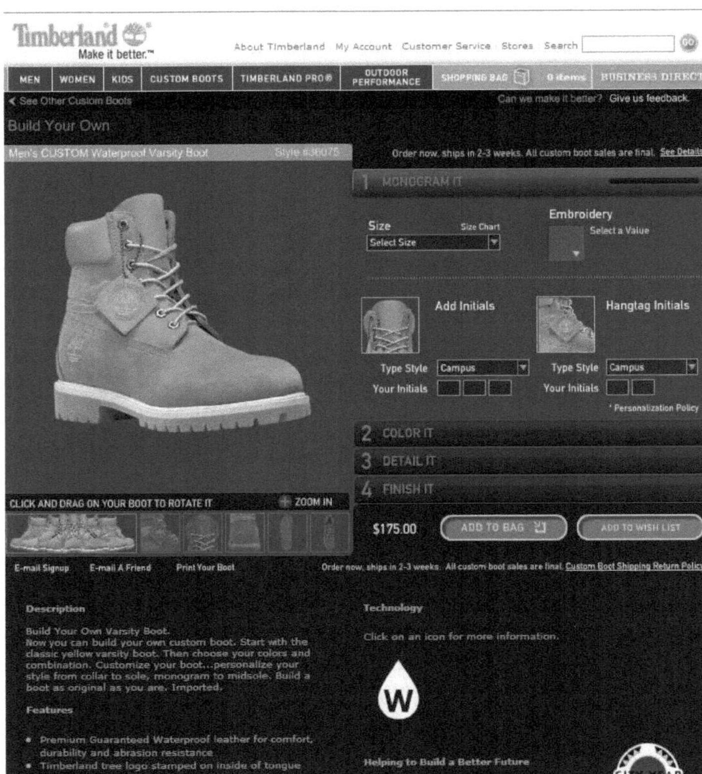

Figure 4.9. The Custom Boot section of the Timberland Website

Timberland included in its new Website a dedicated section with a state-of-the-art configurator, giving many colour options for men's and women's boots (Fig.4.9). The well-known manufacturer of the "yellow boots" now has a specific offer of customized footwear. Accessing the Custom Boot section of the their online shop, it is possible to chose among two basic models (the 6 inch and the

Varsity boots) with a total of 35 variants all together; a very rich catalogue to chose from, with all products apparently sold for the same price of US$175.

Each one of the 35 variants can be individually configured using a very modern, state-of-the-art configurator, full of instructions and information on the details of the process, on how it works and how it satisfies consumers' demands. Up to nine configuration options are possible for each single model, including embroidered monograms of the customers' initials.

The configuration process starts with the selection of the desired size of the boots and it develops in four steps: monogramming, colouring, detailing and finishing. In each one of them colour and materials can be chosen and the result of the configuration choice is instantaneously rendered on the screen, which is a good tool to guide the customer through a rather complicated selection process and at the same time educate him or her in making the appropriate choices.

Timberland declares that it is able to deliver the customized shoes in two to three weeks, which is somehow better than what its competitors seem to be able to do; a long text section describes the modalities for the delivery and the return policy adopted by the company. Timberland has since long completely outsourced its production, so it is likely that the two families of customized boots are produced in the same facilities that manufacture the other models of their collection; we have no indications that the same product offer is (yet) available in the various Timberland shops around the world.

7. **JG Customs** [date of survey: November 2006]

CUSTOMIZATION LEVEL	Style-aesthetic; shoes provided by clients
ORIGIN OF THE COMPANY	innovative start-up
PRODUCT TYPE	Sneakers
PRODUCT POSITIONING	unique business model; top market
OPERATIONAL APPROACH	pure click, but with craftsman abilities

The case of JG Customs is difficult to classify following the framework that we have introduced in the previous sections; they are not producers of shoes and not even strictly resellers: they could be called "customization brokers" in the sense that they help consumers to create their own personalized pair of shoes (Fig.4.10).

What JG customs basically offers its customers is a "hand painting" service for their shoes. Clients can provide JG with shoes they bought themselves or they can ask the company to purchase the shoes for them (there is a list of different models of sneakers of various brands they can work on). Once the shoes are available, the company actually repaints them, using special colours and techniques, following the specifications and the desires of their clients. They mediate between the creative demands of the consumers and their lack of ability to work on their shoes themselves[33], by selling their skill on one hand and creativity on the other.

[33] But to those who think they are good enough to do the job themselves, JG Customs offers a "self-customization kit" with the tools and colours that are necessary to paint the shoes.

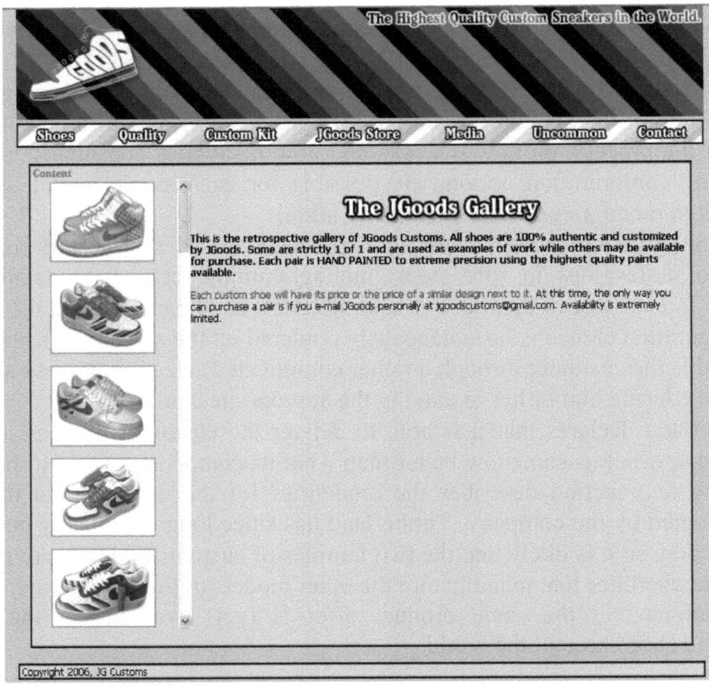

Figure 4.10. Some of the "Creation" of JG Customs

This kind of unique service offering is, in any case, not for everybody: the cost for the customized shoes is around US$500 to which the purchase price of the original sneakers must be added; and the delivery time of the hand-painted shoes is two to three weeks. Not a mass market product, but rather an interesting niche for the fashion or design addicted; JG Customs seems to have been able to catch their products demand.

8. O'Neill– design your sneaker [date of survey: November 2006]

CUSTOMIZATION LEVEL	Style-aesthetic
ORIGIN OF THE COMPANY	innovative shoe maker; global brand
PRODUCT TYPE	Sneakers
PRODUCT POSITIONING	no real products for sale
OPERATIONAL APPROACH	pure click; no shops; no production

O'Neill's excursion into product personalization is more an open innovation experiment with their customers than a true mass customization project. Nevertheless it is worth mentioning here because it is a good example of the involvement of the consumer in the product life cycle.

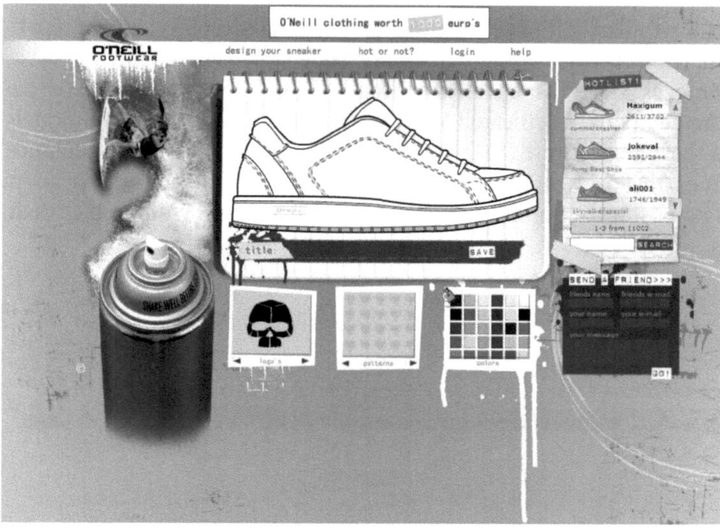

Figure 4.11. The O'Neill design your sneaker Website

Consumers who access the Website are not buying any shoe; they are offered the possibility of designing (more precisely colouring) their personal version of a typical O'Neill's sneaker (Fig.4.11). They can satisfy in this way their creative talent and participate in a sort of worldwide design competition (at the date of this survey more than 10,000 designs had been submitted from all over the world), in which all proposed models are evaluated in terms of "coolness" by the vast audience of Web visitors.

The motivations for customers include enjoyment and fulfilment of personal tastes and interests, and comparing one's talent with many others for no cost and little effort. But for the producer, the returns on the investments for this initiative are much more important; what they are running is in fact an experiment in the co-creation of new styles with their customers, as if they were involving a vast panel of consumers in their design team. A great value for the money spent in the implementation of the infrastructure. It would be no surprise if in the near future O'Neill decide to give their clients the possibility of actually buying and receiving at home the shoes they have designed.

9. Footjoy – Myjoys [date of survey: November 2006]

CUSTOMIZATION LEVEL	style-aesthetic; size declared by consumers
ORIGIN OF THE COMPANY	innovative shoe maker; renown brand
PRODUCT TYPE	high quality golf shoes
PRODUCT POSITIONING	true mass customizer
OPERATIONAL APPROACH	pure click; no shops; outsourced production

Footjoy is a popular golf shoe manufacturer that has recently started to offer a line of customized golf shoes through its online shoe store (Fig.4.12).

The process starts as usual with the selection of the model to customize, through the company's dedicated Website (Myjoys). There are five models on sale for prices that range from US$140 to US$190 per pair. The more technical nature of the shoes that are offered (golf shoes) compelled the producer to put more emphasis on customization parameters such as length and width rather than on pure aesthetics.

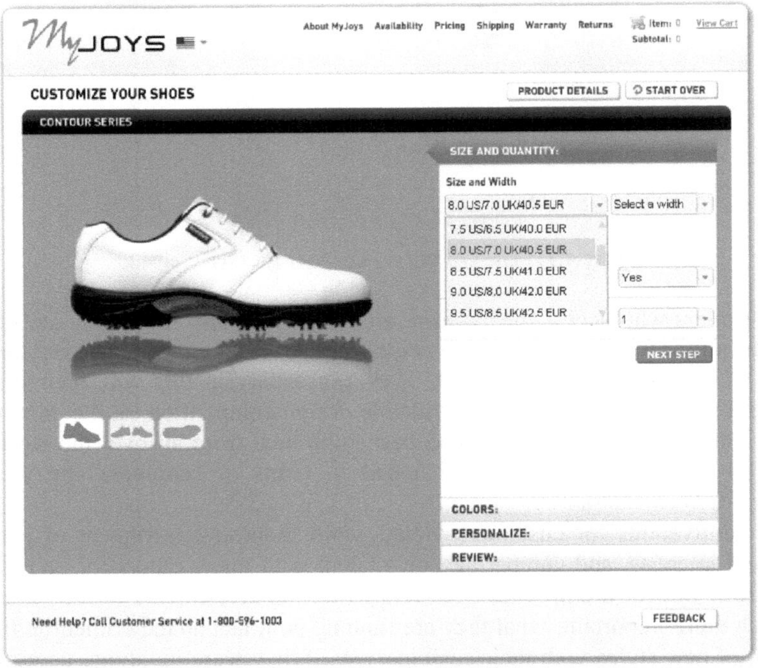

Figure 4.12. The starting page of the Myjoys configuration process

In fact Footjoy is the only company that lets the consumer choose between a high number of lengths and four different widths (from narrow to extra wide); in addition to that left and right shoes can have different values of length and width. Quite a high choice, with the only drawback that consumers are supposed to know all these minute details about their feet; and this does not happen so often. The next steps are very similar to those of all the other style customizers: base, saddle and accent colours can be chosen from a rich palette of colour shades

To complete the personalized design a monogram and logo (with options that range from skull bones to national flags) can be embroidered on the upper. A rich enough choice to satisfy the aesthetics of golf playing, combined with a much more substantial and valuable selection in terms of fit combination. It would be perfect if combined with an actual foot measurement to identify and select the best combination of values to fit the consumer's foot, but this doesn't appear to be the strategy of Footjoy (yet), since these shoes can only be ordered online in the Myjoys Website; there is no evidence of such sales in real shops.

In this final section of the chapter we have included the most recent "newcomers" which have only recently started offering mass customized shoes.

10. Steve Madden – Design your own collection [date of survey: November 2006]

LEVEL OF CUSTOMIZATION	Style-aesthetic; size declared by consumers
ORIGIN OF THE COMPANY	manufacturer of trendy women's shoes
PRODUCT TYPE	high quality and fashionable women's shoes
PRODUCT POSITIONING	true mass customizer
OPERATIONAL APPROACH	pure click offer; outsourced production

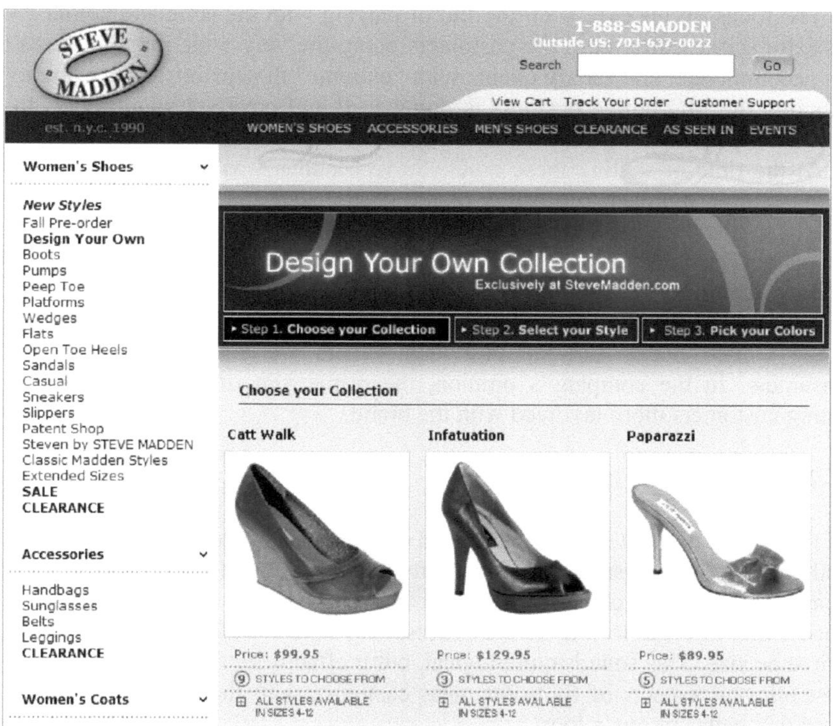

Figure 4.13. The Website of Steve Madden

Steve Madden is a New York-based manufacturer of trendy women's shoes. In February 2006, the company became the first women's dress shoe company to offer its customers the chance to design their own shoes over the Internet (Fig.4.13).

The approach chosen by Steve Madden in its implementation of mass customization is very similar to the ones adopted by the big sports shoe brands: style customization via the Web, through a dedicated section of the company's well presented Website (called "Design your own collection"). If we put aside the diversity in products (women's dress shoes rather than sneakers or sports shoes), the steps of the Web-based process are very much the same: there are three models on sale (Paparazzi, Infatuation and Cat Walk), for prices that range from little less than US$90 to a maximum of US$130. For each one of the basic models, several preconfigured styles (DYO – design your own Steve Madden) are available. Once selected, the basic style can be further personalized, by choosing the colour of the upper and of the heel; the choice of colours is very rich and many different combinations are possible. The selection of the size (the company offers a much wider range of sizes than normally available in shops) concludes the configuration process and the transaction. A very neat and simple setup, which is worth being analyzed for the fact that it applies to women's dress shoes (much more style and fashion based) what can consider to be a consolidated customization path in the segment of sport shoes.

The focus in this case is on the fun of playing with the style and colours, with the extra service offered to the customers being the very wide range of sizes that can accommodate and satisfy clients with "extreme" dimensions. According to the company the MC online business is doing well and positively contributes to the global company Internet sales.

At the time of writing these notes, Steve Madden was still on a "pure click" customization approach (no offers of the customized shoes in shops); the shoes ordered by the customers are manufactured in China by the normal suppliers of the company. The long distance from the outlet markets is not considered to be a limiting factor, as long as professional carriers can guarantee fast and reliable deliveries. Shoes are transported in bulk from China, then shipments are broken up in the United States and the individual orders are delivered directly to the consumers. In the company's opinion, the prime aim of this experiment is in getting customers more involved with the brand.

4.3.2 Best-matched Fit

All the companies that have so far started offering customized shoes privileging the aesthetic vector of mass customization are mainly using online sales in the Internet as a preferred sales channel. Whatever the approach is and the details of the process are, in order to initiate the process, customers are case supposed to know their shoe size. This sounds quite logical, but it effectively risks spoiling one of the most relevant aspects of footwear mass customization: the possibility of buying shoes that better fit one's foot.

The second group of footwear mass customizers that we are going to deal with in this chapter have decided to fully exploit this possibility, assuming (and surveys on consumer interests have proved that in many occasions) that their customers would have been much more attracted by this possibility than by the simple option of playing around with the style configuration of their shoes.

Hence, the question these entrepreneurs are trying to answer is: can shoes be made in such a way that they perfectly fit the morphology and the dimensions of anyone's feet? We already know the answer that this question received in the past (and still today) by craftsmen and cobblers: it is certainly possible by carefully measuring the feet (left and right individually) of the client, by manually shaping a pair of lasts that perfectly fit the foot and by once more manually constructing the shoes and all their components. It can be done and is done for all those customers than can wait several weeks for their shoes and who can afford the huge amount of money this way of doing things costs.

In order to achieve the goal of mass customization (which is the one of making personalized products affordable by the majority of the consumers), different methods and technologies must be used and simplifications must be introduced in the process, to obtain a compromise between the level of comfort that can be granted to the consumers and the complexity (and hence the cost) of the whole sales, design and manufacturing processes. One easy way to simplify the process and keep costs at a reasonable level would be by avoiding the production of tailor made, individually adjusted lasts. If this is done, the producer can still take advantage of the modularity (components and tools developed – graded – in a set of proportional dimensions) of the classical shoe fabrication process and of the economies of scale of this kind of industrial production.

This way of accomplishing fit customization is called "best-matched fit" (see also Chapter 2); what happens in practice is that foot dimensions are used to search a database of lasts graded in all the necessary combinations of length and width, choosing the one that is the best match of the customer's foot.

Clearly this is not like making true tailor made lasts and shoes, but does it ensure that consumers feel more comfortable in their shoes? We are convinced that this is the case. We know that it is neither easy nor consistent from shoe to shoe to tell and consumers most of the time have to learn from their mistakes. Having an objective, rigorous way to measure foot dimensions (using foot scanners), and using such dimensions to chose the most appropriate shoe model or shoe size, is enough to guarantee a "quantum leap" in terms of perceived comfort and customer satisfaction. If, as an extra, mismatched pairs are offered (left and right shoes different) the level of comfort which the consumer experiences is very close to that of a tailor made shoe, yet at a cost that makes it affordable to vast segments of the population.

There are five examples of mass customizers of this kind that we are going to document in this section and the next: the American company Otabo, the Finnish LeftFoot and the German Selve, Viavor and the Slovenian Alpina. Despite the different kinds of shoes they offer and the different details in the way the sales process is implemented, they all have several aspects in common: they sell through real shops, they use foot scanners to take consumers' feet dimensions, and they rely on the Internet as a parallel and complementary sale channel; but customers have to go first to their shops. In addition both Otabo and LeftFoot have their own manufacturing facilities.

1. Otabo [date of survey: November 2006]

CUSTOMIZATION LEVEL	fit and style-aesthetic; feet measured with scanners; best-matched fit approach
ORIGIN OF THE COMPANY	start-up with new brand
PRODUCT TYPE	upper class men's dress shoes
PRODUCT POSITIONING	affordable luxury
OPERATIONAL APPROACH	click and brick; real shops and owned manufacturing facilities

Otabo (Fig.4.14) is a (relatively) young and upcoming US brand and manufacturer for mass customized men's shoes, with a growing number of stores all over the States. Its shoes are manufactured in the USA (in Florida). The history of the company is a history of determination, faith in the market and of huge investment in technology, at all levels.

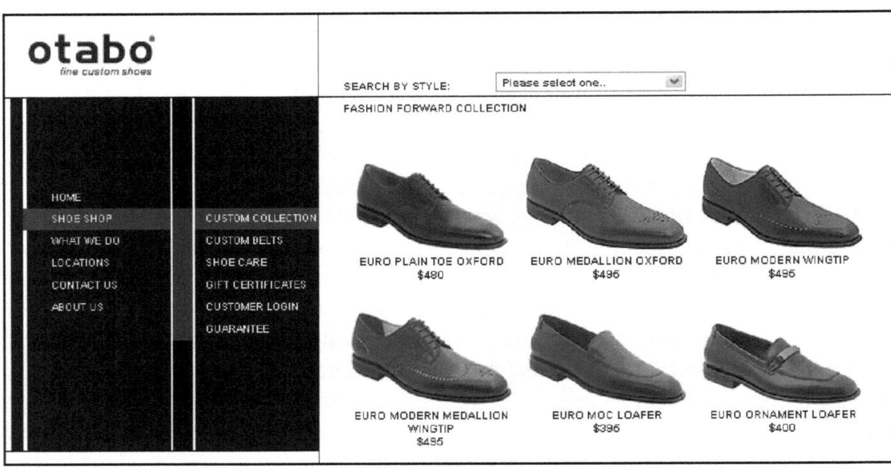

Figure 4.14. A section of the Otabo Website

The founder and owner of Otabo, started working on this idea more than ten years ago; it took him time and a huge amount of money to chose and develop the appropriate technologies both to support the sale of its customized shoes and, more impotantly, to be able to produce them with the quality, flexibility and reliability needed. At the time of writing these pages the company runs ten shops (all in the United States), with more are on the verge of being opened, and offers a very rich collection of casual and dress men's shoes for a cost per pair that ranges from US$350 to US$450.

As with all the other companies in this section, the first approach with Otabo is to go to one of the company shops in the various cities of the United States. After the foot scanning step (which is done by a state-of-the-art automatic foot scanner), the client is guided through the shoes configuration process.

The collection is very extensive and there are many details of the shoe that can be customized according to the desires of the consumers: of course the colours and the materials for the various part of the upper, the colours and materials of the outsole, the shape of the toe and the style and design of the decorations on the upper. Orders taken by the various shops are forwarded to the manufacturing facility in Florida, where the shoes are manufactured using a highly automated process using modern and highly flexible shoe machines. Foot data are processed there, using CAD systems (widely used also in the shoe design process) to identify the optimal last size for the client and to generate all the data that will be later needed at the manufacturing level.

The high level of automation and of digital control on the various process steps allows Otabo to run its factory with a limited number of highly skilled workers; this, together with the high degree of vertical integration of such processes (the place where the factory is installed did not have any network of suppliers available; so Otabo manufactures its own lasts and outsoles within the factory with a level of vertical integration that is impossible to find in traditional shoe companies nowadays), enables the company to be competitive on manufacturing costs even if they are based in a high labour cost country like the USA. When ready, shoes are dispatched either to the shops where the customer ordered them or directly to his address.

A quick look at their Website gives a clear impression that the key assets of the company are its products, its shops and the control on the entire design and manufacturing process; the Web is more a window to promote the company and its collection than the place where real sales are made. Customers of Otabo, having gone through the scanning process and having received their ID number, can of course use the web for new orders, but in this case the "click" is a mere support and complement to a very much "brick" orientated company strategy.

2. **Viavor** [date of survey: November 2006]

CUSTOMIZATION LEVEL	fit and style-aesthetic; feet measured with scanners; best-matched fit and custom fit approach
ORIGIN OF THE COMPANY	spin off of technology supplier
PRODUCT TYPE	men dress shoes
PRODUCT POSITIONING	affordable luxury/mass customizer
OPERATIONAL APPROACH	pure brick; only real shops with manufacturing outsourced

Viavor started as an initiative of the Canadian company Vorum, a well-known provider of foot scanning technologies and of software and hardware solutions for orthopaedic shoe producers. Some years ago, Vorum realized the market potential of mass customized shoes and it felt to be in a favourable position to exploit that market, since they controlled most of the more relevant enabling technologies for such a process. Vorum than created Viavor, which is now offering directly and indirectly (through a network of resellers and franchisees) mass customized shoes to customers all over the world.

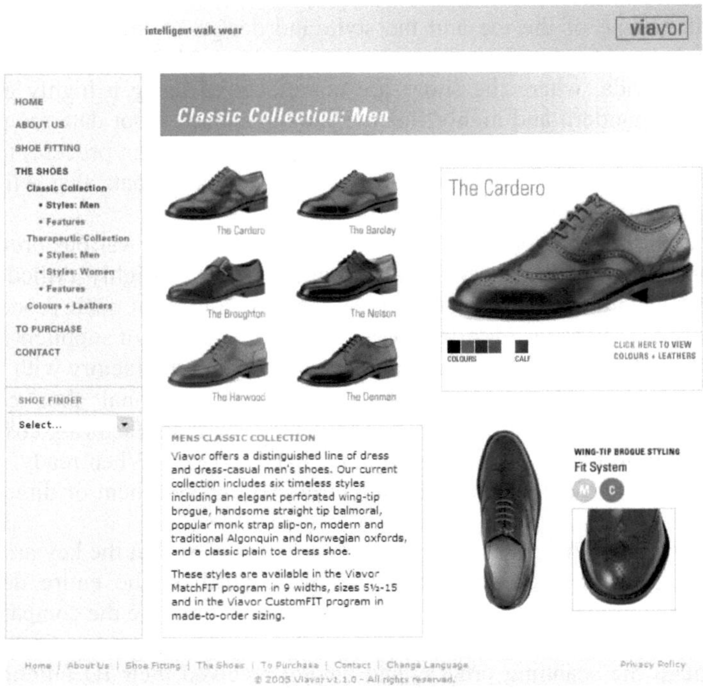

Figure 4.15. The Viavor offering of customized men's shoes

The Viavor case is somehow different and peculiar with respect to others that we have covered in this study; to begin with, the Website of the company (Fig.4.15) rather than selling shoes (no Web sales are offered to customers) carefully illustrates how the customization process happens in real shops, which is the only channel Viavor shoes are offered from. What strikes the observers of the Viavor Website is that there is no emphasis on style and aesthetic customization; the message the company wants to send from its Website is all about the substance of the process: *fit customization*. The three steps of the process are carefully explained: the feet are first measured (using the well-known and proprietary Vorum technology); then foot size and shoe (or better, last) size are compared and matched (to identify the optimal last for the customers, selecting it from a data base of last graded into 20 size and 9 fit-widths for each size); the process is then concluded by a fit confirmation, *i.e.* by a try-on test in which the customer tries some sample shoes available in the shops in order to verify the selection made by the software. As can be argued, this is a very standard approach to fit customization, using foot scanners in a real shop. Vavor insists in the details of this process to signify how important and relevant they are to the correct selection of the shoe.

In addition to that, Viavor seems to be the only mass customization company capable of offering true custom fit (bespoke shoes). As the Company declares:

> *"If a good match is not available in all the sizes provided in the MatchFIT program, or for those customers who prefer a fully custom made shoe, Viavor also offers a traditional made to order service (of custom made shoes) - with a high tech advantage!" (from Viavor's website)*

In this case a personalized shoe last is designed using the advanced software tools developed by Vorum to design and produce orthopaedic shoes (which are by definition entirely individually fitted) and all shoe components are specifically manufactured on the personalized lasts. The experience of the company in the orthopaedic field is clearly evident from the fact that besides the line of customized men's shoes (the MatchFIT and CustomFIT Classic), a second line of CustomFIT Therapeutic (both for men and women) is offered.

Six basic men's styles are offered, all with a classical and highly valuable Goodyear construction; three sole options are available (all leather, leather and rubber, and all rubber), together with a good choice of colours, materials and other style details. The Website looks more an online catalogue than a true configurator (also for the fact that it is not possible to order shoes over the Internet), but it gives an idea of a rich enough collection of models, all very elegant and with an expensive and classic construction. Once more it is clearly evident that the value Viavor is offering is in fit customization (including custom fit), with style only coming as second.

As far as the processes are concerned, Viavor is pure brick: its shoes are only sold in real shops throughout the world (three in Canada, two in the United States and three in Germany). Surprisingly enough, Viavor does not control manufacturing but instead completely outsources its production to China where all its shoes are made with the required quality and reliability in deliveries. Here again, the distance of the manufacturing units from the outlet markets does not seem to be a limiting factor; a strategy exactly opposite to the one adopted by Otabo. Although no precise price indications for the Viavor shoes could be gathered, the impression is that they will be positioned (at least price wise) at the limit of the luxury segment, more towards the lower segments of the price pyramid, which justifies the mixed definition of affordable luxury and true mass customizers.

3. **Alpina** [date of survey: November 2006]

CUSTOMIZATION LEVEL	fit and style-aesthetic; feet measured with scanners; best-matched fit approach
ORIGIN OF THE COMPANY	traditional shoe manufacturer
PRODUCT TYPE	mountain and trekking boots
PRODUCT POSITIONING	mass customizer
OPERATIONAL APPROACH	pure brick; no Web sales; own manufacturing

Alpina is a traditional producer of ski, mountain and trekking boots as well as men's and women's dress shoes, based in Slovenia; a few years ago Alpina has launched an experimental program in the field of mass customization, installing foot scanners to select and order customized mountain boots.in a large number of its shops (Fig.4.16) distributed in many of the former Eastern bloc countries.

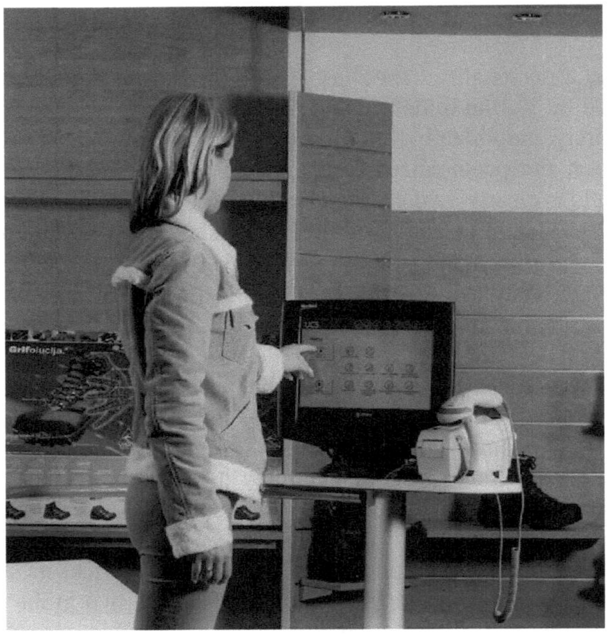

Figure 4.16. The selection process in one of the Alpina shops

We could not gather many details on Alpina, but we thought that the Slovenian manufacturer was worth a mention. As in the case of LeftFoot (which will be presented on the next page), we have a traditional shoe manufacturer, based in a very stable consumer market, seeking added value that it can offer to its customers. Alpina was convinced that driving consumers in the selection of the appropriate size for their mountain and trekking boots would have been a relevant added value to induce them to stay with the brand and to become more and more fidelized to it.

Foot scanners are installed in the shops and through a very simple process, the basic foot measures are obtained; these foot measures are used to search in the last database for the best matching one (given the shoe model that the customer has previously indicated) to suggest the optimal size/fit combination to match the consumer foot. As in the case of Viavor, shoes to try on are available in the shop in order to confirm the choice of the software prior to placing a final order to the factory. A simplified configuration process is also done in the shops to select the desired colour/material combination for the shoes. Manufacturing is handled by the Alpina plants located in Slovenia, very close to the destination markets of the customized shoes.

4.3.3 Detailed Case Studies

4. **LeftFoot** [date of survey: November 2006]

CUSTOMIZATION LEVEL	fit and style-aesthetic; feet measured with scanners; best-matched fit approach
ORIGIN OF THE COMPANY	new brand of a well established shoe company
PRODUCT TYPE	high segment men's dress shoes
PRODUCT POSITIONING	true mass customizer-affordable luxury
OPERATIONAL APPROACH	click and brick; real shops and owned manufacturing facilities

LeftFoot is a leading European provider of custom footwear for men, with stores all over Europe and headquarter in Finland.

Figure 4.17. A section of the LeftFoot Website

Company profile

Established in Pomarkku, a Finnish city near the coast whose footwear heritage is still important, nearly 50 years ago, Pomarfin has become Finland's leading manufacturer of casual footwear for women, men and children; its LeftFoot branch specializes in products for men only. Pomarfin was not hurt so much by the crisis of the Finnish footwear industries of the end of the 1980s. In the first half of that decade, Finland exported its footwear production to the Soviet Union (nearly 40% of production in the first half of the 1980s), but after the crisis of the break up of the USSR, the market closed its frontiers and Finnish shoe production had to face severe problems.

The worst years for Pomarfin were 1995/96 when the company had heavy financial losses. A direct consequence of this crisis was the decision to find a new line of products with a new market: the luxury market. Pomarfin then moved all its production to Pärnu, Estonia, in order to cut labour costs. The challenge was in terms of setting up a new production line in a new location, which means different people to be trained in new production systems. This is the brief description from the corporate LeftFoot Website (Fig.4.17):

> *You can get a pair of made-to-measure Left shoes manufactured from high quality materials that you have selected, delivered to your door. Once your feet have been scanned at a LeftFoot company outlet and you have signed in to our system you can order new shoes via our web pages.*

The Move to Customized Products: When and Why

Pomarfin integrated customized products with classical production in the middle of 1980s as an effective answer to the crisis of the Finnish shoe market caused by the downfall of the USSR empire. Facing price competition Pomarfin worked with innovative materials - Gore-Tex® lining and the Fenomex® sole - presenting different options to the customer. Unfortunately, the volume of the Gore-Tex® shoe market was not so large as to become a successful choice. In 1996 the company made a market survey in Finland to identify customers' preferences and new market trends, and the result was the demand for customization and personalized products by creating individualized shoes.

Some time before the realization of the questionnaire, the executives of the company had already discussed the possibility of manufacturing shoes made to order. The idea itself was not new, but the matter was to have personalized shoes produced on a mass scale, which implied a different production approach, and the company was not prepared for that. Moreover, the company wanted to create additional customer value and to fulfil existing demands from the market.

Pomarfin set clear goals for the new mass customized footwear concept: the main objective was the creation of a true customer-centred product achieved by controlling the entire value chain. The production is on-demand, and footwear are produced on the basis of existing orders answering customers' demands for individuality and independence. The technical goal was centred on the ability to exploit the economies of scale of mass production.

Customized Products and Marketing

Products are classic shoes for men. LeftFoot divides the collection into three lines:

- business
- fashion
- casual

A relatively new line is the one dedicated to golf shoes. The customer can chose from 700 shoes and 30 models with variations of colour, leather and outsole. For each standard size (from 39 to 48), there are five different width measurements.

The collection changes three times a year. New models are introduced and some of the old models are suspended, but there are some classic models that remain from year to year because the clients are attached to them, and they continue to choose the same style with different soles or different leather suited for different seasons and occasions. LeftFoot has outsourced the original design process to an Italian designer, who creates the designs and makes suggestions regarding the materials.

The customer can choose the outsole, in material and aspect, depending on the different use and weather conditions of the environment, and afterwards can decide the colour and the material of the product. Each single model has a predefined customization range (normally five different leather/colour choices and various soles). An added value is the new golf collection, which has been created for the Finnish market. Golf is the favourite sport of professionals, the target client of LeftFoot. The golf collection doesn't change with the seasons.

To complement this, there's a line of products to take care of the shoe: a polish cream tested for the selected type of leather, a set of brushes and a shoe tree are offered to the client as a sort of service for his precious personal product. This set of products strengthens the concept of pampering the customer.

The Customization Process

The new customer has to physically visit one of the LeftFoot shops (located in several European countries, mainly Finland, Germany, England, Italy but expanding every year). In the shop the first stage is the foot scanning. A scanner instrument registers the profile of the customer's feet and measures them in width and length. Software called PEDUS3D determines which shoe size is the best one for the customer (best-matched fit approach).

Since the customer normally prefers to try on the selected shoes, and to verify the size suggested by the software, the second step is to test a sample pair of shoes, taken from a stock of samples, in all possible models, lengths and widths. Sometimes the customer has differently sized left and right, making it necessary to choose two different measures. The trying-on stage during the first order is important to check and confirm scanning data even if the appearance of the model is not the same as the chosen one. At this stage the customer can choose his pair of shoes. Sometimes he has already decided on the model (because he went on the Website, or he saw it in the shop window), in some other cases he needs the shop assistant's help. The assistant addresses the customer's needs with a few questions about the use of the footwear. Speaking with the assistant the customer can be guided to choose the outsole, the colour and, consequently, the shoe material.

One model can be ordered, on average, in five or six different kinds of leather. Moreover, a Gore-Tex® lining can be added to some models. The customer receives an order confirmation and a customer ID for the Internet service. The customer ID can be used for ordering new models in the future but also to check new collections or new varieties of choices in model. The entire process takes between 5 and 25 minutes.

Customer's data are printed on the insock of the shoe: name, surname and the personal identification code that is the same to use for future orders via the internet. Season's greetings for Christmas presents or personal messages, are an additional customization service.

Once the purchasing is finished, the customer pays the full price of the shoe. The user's data are sent to Estonia for production, and the product will be delivered directly to the user directly in three weeks, with no need to come back to the shop. The customers' information is enterered in the LeftFoot database and remains available for future orders directly from the company Website. The database is a repository of customers' characteristics; it is important for the company to know its target consumers and the best selling products, and allows it to produce focused promotional campaigns. The orders via the internet follow the same procedure, eliminating the personal visit to a shop. The customization process ca be summarized in the following steps:

1. Foot scanning, storing of foot data, detection of foot problems.
2. Trying-on the sample shoe for data checking.
3. Selection of the model - first colour, then material and leather.
4. Selection of the outsole (depending on the customer's needs).
5. Checkout and payment (full cost).
6. Order and data sent to Estonia via Web for production.
7. Posting to the specified address.

In 2003, the proportion of Internet sales was 15% of the total and the proportion is growing continuously. Depending on the market of the country, the on line trade numbers could be important. In Denmark, for example, the quantity of on line buyers is quite large. The explanation of this trend is the fact that the customers are international, people who travel a lot and pass through the city and therefore re-order online from home. Clients who continue to order on the Internet are categorized as satisfied and devoted to the brand and to the product. The store that originally takes the foot measurements "owns" the customer and receives a royalty from every purchase made later on, through the Internet.

Every year the company produces five or six advertisements and information campaigns, targeted to specific customer segments. The campaign normally immediately results in further sales, mostly via the Internet. Customers are contacted personally four times a year, using a newsletter to keep them informed about the launch of new collections or other company events. All advertising is designed for the international market and is translated according to the market area and the country in which it is posted. The shops are in different parts of the world with so many different characteristics that the retailer is a perfect filter to indicate the methods to be used in the advertisement. Retailers have to contribute to marketing and advertising expenses. It should be noted that a lot of work is done at the moment to increase the awareness of mass customized shoes and get potential customers to try LeftFoot shoes.

Production System

The orders are sent to the factory in the evening at the end of each working day. The next day, these orders are processed and go into production. Customized orders are just one separate pair of the ten pairs of each normal production lot.

An insock with the customer's name is printed and used as identification for each personalized pair of shoes, even though the differentiation with the normal passages into the production line is only in the last stages of production.

Materials are purchased from different suppliers; at this stage the same production problems apply as for the traditional manufacturing: to predict the correct quantity of leather for each colour and feature. LeftFoot prefers to have a few trusted suppliers and robust partnerships, because the quality of the leather is important matter. The supplier delivers the leather six to twelve weeks after the order is placed.

This means that knowing the number of previously purchased shoes of each model is paramount in forcasting the amount of each leather required. Soles and other components are standardized and can thus be bought freely. Shoe manufacturing starts when the production personnel cut from leather the upper and the lining, following the order received from the shop or directly from an Internet customer.

Manufacturing uses NC cutting machines that are programmed, based on the customer's measurements and data from the PEDUS3D. The second stage is the stitching, where workers stitch the upper sections together. The sections pass through traditional pre-treatments such as splitting and skiving at the edges, interlining, and eyelet reinforcement. In the next step, the customized Left shoes are dressed on the last, stretching the leather on the selected last, in the size and the model of the chosen shoe.

The shoes are then given a full assortment of traditional heat and vaporization treatments. All these steps are realized to obtain a good quality footwear for size, quality of leather and stitching and aesthetics. A checking procedure is used to test all materials up to the outsole stage and to verify if the outsole is correctly cemented and has a well-finished and good appearance. Tests are not finished at this stage, however, a final control is aimed at checking the total footwear quality in terms of leather, stitching, seams and outsoling.

The shoes are eventually polished, brushed and shined for a final good appearance. Even if the shoe manufacturing (classical and customized) is today a complicated system of machineries, many steps are made by specialized workers who are deeply focused on their duty. Future development in production systems are quite difficult as the process is standardized; there could be improvements, but no revolutions are possible. The finishing stages are then as follows: lacing, LeftFoot logo application (there's a little logo on the left shoe lace), flannel shoe bag, and, if it's part of the customer's order, a shoe care package (cream for the specific pair of shoes and brushes) and shoetrees. The shoes are wrapped in paper and placed into their box. The package is not anonymous, but has the LeftFoot colours and logo. On the shoe box the customer's address is printed and shoes are sent all over the world by the Finnish Post service (except in Japan, where DHL is used). The customized shoes have passed through 40 production steps over six or seven days. When the client receives the shoes three weeks have passed after the order. In conclusion, product development capabilities are important, but not vital for the LeftFoot company. The company has to be aware of what is happening in shoe technology, even if major innovations are unlikely. The process developments on the other hand are imperative, and this is the sector for future progresses.

The Customers' Perspective

There are three possible points of contact with LeftFoot and its customized shoes:

- from articles in magazines
- from the Internet
- from passing by one of the shops

LeftFoot company already has thousands of customers. In Helsinki, for example, the number of re-ordering customers has outgrown the total number of new customers. It is normal for customers to buy two or three pairs of shoes at once. Some customers even buy as many as seven pairs of shoes at the same time. Normally a customer buys two or three pairs of shoes in the shop and later orders more on the Internet.

From a customer's perspective LeftFoot customized shoes are a good product because they satisfy the needs of professionals who don't have time for shopping, for the following reasons:

- after first scanning/visit to the shop they can remain at home buying the shoes using the Internet.
- Men usually love technology, and buying innovative and technologically advanced products is something that stimulates their enthusiasm.
- The customer doesn't need to go shopping - usually men don't think shopping is an amazing activity (women have the opposite opinion). Most often professionals travel a lot for work reasons and don't want to waste free time in shopping.
- Professionals always need a classic pair of shoes and the design of these collections doesn't change so much, so it's easy for them to select the model, even from the Internet. Sometimes they choose the same model in different colours, because of the certainty of the comfort.
- Despite the easiness of Internet shopping and payment by credit card, it's important that the first selection is done in the shop. LeftFoot needs to do the scanning, but also psychologically the customer needs to see the assistant, and see and touch the collection first. The producer has to gain the trust of its new client, which is the reason why in all the LeftFoot shops in the world, even if the products are for men, the assistants are women. This fidelity is the reason of the high success of the Web sales: it is difficult to present all the materials online in such a way that customers can really know what they are going to get when they receive the shoes at home; so shops and retail outlets are very important in the company strategy.

Case Assessment

An important aspect of the LeftFoot Company is the evolution of a traditional company into a new aspect of the product. They applied their knowledge in footwear production to the concept of customized shoes. The cost of the customized shoes is no higher than 10~30% above the normal price; this is because there's no need for stocks and no discounted sales for selling outdated models. An important complementary strategy is the use of the Internet as a privileged selling channel.

Strengths:

- Consolidated company with a long tradition
- Market ready to welcome new models
- No stocks in shops
- No seasonal sales
- No unsold pairs

Weaknesses:

- Not for people with feet diseases
- Shops only in northern countries
- Waiting time for delivering

Opportunities:

- Ladies' shoes, even if the process characteristics will change a lot
- Large potential markets empowered by Internet
- Franchising as a way to expand in different markets
- The use of other shoe manufacturers

Threats:

- A new brand proposes MC footwear
- Minor interest in customized shoes
- Customer perplexity in paying extra costs for the MC product

Future Development

The LeftFoot concept has been based on shoe manufacturing knowledge since the company has a strong background in shoe production, but for the future the goal is to become a system and concept and service provider. The development of the concept and the system in different stages of the process of customization is foreseen: improvements in the measurement software, and an improvement in the service level in order to serve customers better.

An interesting aspect is the training: "teaching" the customer during the buying process, and later on through different forms of customer contact, *e.g.* e-mails, newsletters, customer brochures, customer events, and so on, linking the mere advertisement with effective information about the customization benefits and philosophy. The standardization of foot measurements is a difficult task. The Mondo Point system was an effort to standardize measurements, but it was not successful. Follow-up changes in foot measurements are therefore made separately at each company.

5. **Selve** [date of survey: November 2006]

CUSTOMIZATION LEVEL	fit and style-aesthetic; feet measured with scanners; best-matched fit approach
ORIGIN OF THE COMPANY	start-up with own brand
PRODUCT TYPE	upper class, high fashion women's shoes
PRODUCT POSITIONING	affordable luxury
OPERATIONAL APPROACH	click and brick; real shops; manufacturing outsourced

Selve is a Munich-based custom footwear company for women's shoes. Latest design and best-matched fit (based on foot scans). Stores in Munich and London and manufacturing in Italy.

Company profile

Selve is a young company that offers women a full range of personalized shoes (Fig.4.18). Using a foot measuring device, both feet of the customer are measured; the client can then choose a style in a collection composed by a number of classical and fashionable sandals and shoes. All the footwear can be personalized choosing some variants: material(s), colours, outsole, heel, insole and lining.

Selve currently has two shops, one in Munich (Germany) and the other in London. Both shops have foot measuring equipment, a collection of models with most of the currently fashionable colours displayed, and a storage zone with shoes in different sizes to let clients trying-on the models they've selected. The user can test the comfort of the model she has just chosen, but in different colours, and materials than the one of her choice. Selve carried out an investigation into the morphology of the feet of their target customers, from north to south Germany, and determined a range of shoe sizes much wider than the usual one. The company uses the best-matched fit approach, determining the best measurements in terms of length and width and choosing the most appropriate size of last to match them. The core of the offer is a very large collection of models, because of the many possibilities of aesthetic customization; the models are a determined number of classic shoes and sandals and, on the base of the chosen model, the user can configure her shoe.

Passage toCustomized Products: When and Why

In the case of Selve there was no actual move from traditional production to mass customization; since the very beginning the entrepreneurial idea has been based on the exploitation of the concept of mass customization. The start-up company was funded in 2000.

Figure 4.18. The Website of The Shoe Individualizer, Selve

After some years of background studies and analysis, of careful tuning of the last geometries and measurements of the average biometric data of the target population segment, of attentive definition of the characteristics of the models to be launched, the shop in Munich was opened in 2001 and Selve began with its commercial activity.

Customized Products and Marketing

Products are shoes, sandals and boots for women. Besides women's footwear, the shop in Munich also started testing of the sale of men's shoes. There is a classic collection composed of a number of shoes that have remained the same for some years: these are the best selling models that clients regularly buy from season to season. Other models change each season. In the shop, and on the Web, the company also sells an autumn/winter and spring/summer collection. The catalogue is composed of three sections: seasonal collection, classic shoes, and boots. Here is how Selve explains in its Website its sales and marketing approach:

> *Creating a shoe that's unique to you is as easy as ABC...*
> *Firstly, choose your style from the Collection and then add your own design flourished by:*
> *A - Choosing your heel shape*
> *B - Picking and mixing your colours from the sample book*
> *C - Selecting lining and sole options*

Some of the most faithful clients of Selve have known the company since its very start and they have matured a solid trust in the quality of the products and in the reliability of the processes; new customers are constantly attracted through several advertising means. Periodically Selve places advertisements in the local newspapers, and on local TV or cultural channels to “capture” new clients.

The Customization Process

When a customer enters in the shop for the first time, the clerk explains how the process of personalization of shoes works, and the first step is the measurement of the feet. The customer is asked to take off her shoes and to let the assistant measure both her feet. Five measurements are taken:

- width of the foot
- length of the foot
- width of the heel
- height of the big toe
- instep

For boots the clerk also measures the width of the leg at different positions. The data are stored into a database with other information about any possible problem or peculiarity of the client; these are then recalled in any subsequent selection of a new pair of shoes. The customer then chooses a model of shoe among the footwear she can see on the shelves or after a visit on the company Website. In the sales point, all the possible shoes of the collection are visible; women want to see the different colours and different materials that are available for the shoe models they have selected.

The sale assistant can somehow drive the selection: she has the duty of helping to choose the most appropriate footwear for the feet problems of the customer or for the use of the shoe the customer has in mind. Selve also collaborates with a podiatrist who can help to solve specific foot problems with dedicated insoles; he can also suggest one model instead of another should the first not be correct. The client always wants to try-on the shoe to test the comfort; so shoes of the right style and size are normally available in the shop for the test. This stage allows the customer to test the actual comfort of the shoe she is going to buy.

Once the model and the size have been defined, the personalization phase starts. The client and the assistant sit at a table discussing colours, fashion indications and combinations with dresses, depending on the use the customer has in mind. At this stage the following characteristic of the shoe are defined:

- colour
- material (leather, suede, *etc*) of the upper
- lining material
- style and colour of the heel
- style and colour of the sole.

When the configuration done manually without any specific kind of software is complete, the client receives a summary printout of the shoe characteristics she has selected, as a reminder of the choices she has made; she can leave at this stage and come back later to confirm the order or, if she is ready to proceed, she place the order immediately. If the order is confirmed, all the data about the model, the size and the configuration are printed and sent by fax to the manufacturing unit.

All clients are asked to leave a deposit if they decide to order their shoes, which is normally around 40% of the final price; a payment confirmation and a summary note with all the relevant data of the order are then printed and delivered to the customer. At the end of the process, the customer tells the sales assistant how she wants to be contacted when the shoes arrive (by phone, by mail, at a precise time of the day). The shoes normally take three weeks to be delivered, and the boots four weeks, depending on the workload of the manufacturing units in Italy. Delays in delivery are a problem in customer relationships and can endanger the building of trust of the consumers. If the shoes delivered are not considered comfortable by the customers or are not accepted for any other quality reason, Selve sends them back to Italy to correct the defects or to make a new pair. In summary, the customization process consists of the following steps:

- Foot measurement, storing of foot data, detection of problems.
- Model selection.
- Trying-on the sample shoe for checking the correct measurement (the shoe to be tried on is made on the same last as the selected one, but the model could be different).
- Selection of the heel.
- Configuration of the shoe: selection of colour, material, type of leather, lining.
- Payment of a deposit.
- Printing of the customer voucher.
- Order and data are printed and sent to the factory in Italy (where production is done).
- Waiting for delivery (three to four weeks).
- Delivery of the shoes to the original shop for a further product check.

After a first visit to a Selve shop, the customer can chose and configure her pair of shoes on line, because her data are already in the database. However, the sales assistant has a special relationship with the client, taking care of her by mail or telephone. On the Web the entire collection is also presented, and all the possible choices are offered for online configuration; the personalization path is the same as in the shop.

Production System

All Selve shoes are (presently) manufactured by a selected range of suppliers; in the first years of the company these suppliers were based in Germany, but the company soon realized that for their target customers shoes would have to be of a much higher quality, and Italy was a natural choice for that. This turned out to be the correct decision as far as quality of materials and construction was concerned, not certainly in terms of smoothness of the operations, easiness of management and reliability in terms of delivery and, in general, of the service provided. Manufacturing has been so far the limiting factor to the expansion of the company. Selve has been continually looking for a dedicated manufacturing unit, willing to invest in the same business and to exploit the same market, up to the point of specializing its production entirely to the business of mass customization. New formulae ensuring a more direct control over the different manufacturing steps are being explored by the company to further improve the quality of the service provided to their customers.

The Customer's Perspective

The customers are women, aged from 16 to 70 years old, who buy personalized shoes for different reasons, some for aesthetical reasons, others for comfort only. There are three possible points of contact with Selve and its customized shoes:

- by reading an article in a magazine (normally a women's fashion magazine; some examples are on the Website of the company in the "press" section)
- by already knowing the shop or from previous customers (word of mouth)
- by passing the shop - many customers discovered Selve thinking it was a "normal" shoe shop. They entered the shop and asked for the shoes in the window and then realized they could have them any size they wanted.

The first and second are the most usual methods of getting in touch with Selve. According to Selve, their clients approached the company for three main reasons:

- for foot problems and malformation
- for the pleasure of "designing" their own pair of shoes
- for special fashion occasions.

When asked whether the comfort of the design is the most important aspect for them, customers give the same importance to both aesthetics and comfort. Sometimes customers are women of different ages (from teenagers, to middle aged women, to pensioners) who need a pair of shoes for a special occasion, such as a marriage, a particular event and so on. They go to the shop with the dress they want to wear and design the shoes they want in terms of shape and, usually, in a colour to match the dress. The first and main difference to the sales approach of standard shoes and shops is the relationship with the sales personnel and the trust relationship that is built up with the client. The possibility of designing the shoes is certainly relevant to the clients, together with the reliability of the comfort. Some definitions that qualify the buying experiences of the typical Selve customers, in their own words, are as follows:

- special
- individual and personal

- funny
- relationship based (with the sale assistant)
- better knowing what customer really wants
- a pleasant waiting time (like the excitement of a child waiting for Christmas night).

Case Assessment

Strength:
- Young dynamic company
- Academic background, research in footwear sector
- Well-prepared shop personnel and good relationships with customers
- Quality of the design (fashion compliance)

Weakness:
- People prefer to go personally to the shop to choose shoes
- Shops are only in two cities
- Waiting time for delivery
- Critical dependence on non-motivated manufacturers

Opportunities:
- Franchising as a way to expand in various markets/countries
- Alliances with other shoe manufacturers
- Exploiting being the only producers of customized women's shoes

Threats:
- Market entrance of fashion brands
- Expanding into the women's sector of more established MC providers

Future Development
Selve has already two shops, one in Munich and the other one in London; the two cities are quite different in terms of population, economical level (life style) and social life, but the concept of customized shoes is attractive in both environments. The next step in the business model of the company is to find a stable and reliable solution for the production side. An important step is to identify a good producer who can assure good products, using high quality materials and respecting time and quality indicators. The idea that the shoes are produced in Italy gives an added value to the footwear. The perspective to open new shops in different countries is foreseen, yet it is currently limited by the production bottlenecks.

4.3.4 Custom Fit

This section deals with the highest grade of mass customization in terms of the perceived comfort for the consumer: namely, custom fit. Referring to what we stated in the previous sections, best-matched fit relies on a compromise (the ideal last is selected within a range of already defined ones) in order to keep manufacturing and, in general, process costs low; and low process costs mean a cheaper price for the customer.

Custom fit would then mean a last and a shoe that is exactly and strictly tailor made on the dimensions and morphological data of the consumer's foot, with the right shoe different from the left one. Once more, this is exactly what shoe craftsmen shoe (cobblers) used to do in the past and still do for their elite consumers.

Are there examples of this kind in the world of shoe mass customization? And, more precisely, are there companies that are able to individually tailor their shoes, dimensionally and aesthetically, for each single consumer? The answer is "no" from one standpoint and "yes" from another. We will try to clarify this statement in the next few pages.

If we go back to the concepts that were introduced in Chapter 2, we can see there is a major difference in the way processes are structured in the case of best-matched fit and of custom fit.

In the first case (best-matched fit), the last and all shoe components are predefined and, possibly, pre-acquired since all of them are obtained from the so-called "base size" (used to produce the base design of the shoe) with a normal grading process, typical of the traditional shoe development activity. Depending on the number of different size and fit (width) combinations offered, all these components can be kept in stock by the company. When an order for one pair of a given size and width arrives, the shoe is simply "assembled" by picking the appropriate components from the shelves. This is possible when the number of combinations is not so high; a rough limit could typically be 10 sizes and 3 fits per size dimensional range, giving 30 different lasts, insoles, outsoles, heels and so on (in terms of dimensions). Moreover, in such a case, the probability that two or more customers can have the same size and fit is higher (because the range of choice is smaller), hence it is more probable that one can run small production lots of identical size (using, for example, lasts of the same size).

This is possible, but not appropriate: 30 size combinations are not really ideal to best serve consumers; the larger the dimensional range, the more likely it is to get very close to the dimensions of a customer's feet, and the closer the best-matched fit last will be to the volume of his or her feet, which means higher comfort. So a much wider range of choice would be advisable: i.e. ten length sizes, plus half sizes, and five widths would a much more preferable choice. Certainly, the greater the number of sizes, the better. It is clear that moving along this line would take us to a point where it becomes impossible to pre-store lasts and shoe components; instead they should be fabricated on demand. This is the first consideration we would like to highlight here: the more we develop the best-matched fit idea in terms of offered range of sizes, the more it will require that lasts and components are made on demand, exactly as a manufacturer of tailor made shoes would do. A manufacturing scenario that transforms from "push" to "pull" but still retains the modularity typical of the industrial production of footwear.

Opposite to that is the case of true custom fit: the real difference here is that, since lasts are "manually" adjusted (rather than been graded and chosen like in the previous case), all shoe components are manually adjusted too and must be redesigned and manufactured accordingly. This is not only manufacturing on demand but design on demand too.

So, to summarize:

- Best-matched fit, at least for a high range of choice of size and fit means standardized design (no design adaptation) and manufacturing on demand.
- custom fit in all cases means design and manufacturing on demand: each customer order requires an adaptation of the design of the shoe, which must be done before the manufacturing process can start.

With this in mind, we can state that, if we leave aside craftsmen and cobbler (those that we defined at the beginning of this chapter as "digital tailors"), there are no companies in the field of shoe mass customization that are offering true custom fit footwear[34]. That's why we should answer "no" to the question we raised before.

But does this make any real difference for the consumer? We would be tempted to say "no" again. If the range of dimensional choice is wide enough and the grading grid is fine enough, we believe that there is no perceived difference in comfort in going from a tailor made last to a closely matched one. In other words, we can make the following assumption:

- *With an increasing number of size and fit (width) combinations in the range of lasts offered, best-matched fit lasts would very closely approximate the results of tailor made (custom fit) lasts and would guarantee the consumers very similar levels of perceived comfort.*[35] *Custom fit is the "mathematical" limit to which best-matched fit tends with a increasing number of size and fit combinations.*

Then, depending on how wide the range of dimensional choice offered, mass customizers working with the approach of best-matched fit can be regarded as "almost" tailor made producers. That is why a positive answer to the previous question could also be possible. We don't have detailed information to state how close to that definition Otabo, LeftFoot and Selve would fall but we thought it was worth discussing this issue in order to prove how blurred can be the border between best-matched fit and custom fit providers. The only mass customization company which seems capable of producing real custom fit shoes, is apparently (if we have to judge from their statements) Viavor.

But what matters more than anything else is the fact that such a border would not be perceived by the consumer; in other words, we are convinced that the extra complication and extra costs that a true tailoring of the last implies are not worth the increment in perceived comfort that this brings. On the other hand, the wider the range, the more producers are forced to move to an "on demand" fabrication of lasts and components, and so their processes, at least on the side of manufacturing, will resemble those of true custom fit providers. As usual, a trade off will have to

[34] With possibly only the exception of Viavor, as we have already presented.

[35] We assume here that the producer offers in any case the possibility of having different sizes for left and right feet, which gives consumers an enormous increment in comfort.

found between the level of service that has to be provided to the consumers and the complexity of the process itself, which dictates the costs of the final product.

The Luxury Brands

Prior to concluding this section, we need to make a final comment on the personalization service offered by some the most renowned luxury brands; this aspect of customization, possibly yet not addressed to the masses, is worth mentioning and at least "positioned" with respect to the other approaches we have described so far.

We refer here to the "tailor made" service that some major fashion brands (such as like Salvatore Ferragamo or Gucci), or high end shoe makers (such as Bruno Magli, Santoni, Berluti or Testoni in Italy, or Church in England) are offering their customers. Top clients are given the service of having shoes which they can order individually, in the size they desire[36] and with the colours and materials of their choice, albeit within a well-defined range of models and materials and colours. A service of this kind is normally very highly priced on top of a product that is already positioned in the top range in terms of quality and cost.

Although in these examples the notion of customization (shoes made to order and configured following the specifications of the consumers) is certainly preserved, the masses do not emerge here; it is an elite service offered to a very small number of customers with no use of specific technologies and relying on the "flexibility" of manufacturing and of the supply chain; as long as these sales account for a limited number of pairs, their production can somehow live together with the normal one with no major disturbances. Hence there is no real mass customization, but a further indication of how important it can be to follow the individual desires of very select customers. By the way, this is also remerging as a very valued service offered by some of the big apparel brands (like Zegna or Versace for example) that are more and more offering their customers (organizing specific events at their flagship stores) a service of tailor made suits and garments.

4.3.5 Functional Customization

The last case we want to treat in this chapter is functional customization. This is the most complete approach to mass customization we can think of: all the three vectors, style, fit and function are applied to this concept and a full exploitation of the idea of customization happens here.

Whilst aspects such as style and fit have a sort of general value, in the sense that they apply more or less with same notion to shoes of different kind and of diverse construction, function is of a more intricate and specialized nature. It is not accidental that the only example that we can bring here (Adidas) is a sports shoe manufacturer; sports shoes by their nature, or at least sports shoes of a certain level, focus very much on enabling their users to attain the highest level of

---

[36] In some cases the size is declared by the customers, in some others feet are manually measured in the shops; we are not yet aware of any one of these companies using foot scanners for this purpose.

performance. And since athletes and sportsmen are not all the same, it is no surprise that football boots for the top soccer players or shoes for number one tennis performers, are hand made or hand adapted individually for each one of them.

Functional mass customization aims at making the same level of service affordable to normal persons, hobby players, sports enthusiast and the like who simply want to have a better and more enjoyable performance. We will see how this happens in the only significant case of this kind: the MiAdidas project.

Is functional customization applicable only to sports shoes? The answers lies mostly in the notion of "function". What differentiates a simple pair of sneakers that some people wear most of the time, except when taking part in sports, from a pair of running shoes? The answer is in how much attention the manufacturer has put into the functional aspects (rigidity, cushioning, elastic response and so on) rather than on the aesthetic and fashion ones. Once more, from the perspective of a normal consumer, the boundaries between these aspects are perhaps somehow blurred; but when such shoes are used for the purpose they were originally conceived for, these aspects are back in all their importance.

Following this line of reasoning we can think of any type of shoe in term of having functions to be performed: we can think of foot protection, comfort in standing or walking, ensuring a certain level of climatic comfort. And this applies at various levels for any type of shoes. But it is enough to think of footwear for specific ages (like for babies and children, or for the elderly) to understand how much function can (or should) prevail on other aspects, not to mention specific less fortunate groups of persons such as diabetics. And the more function becomes important, the more tuning this function to the individuality of each consumer can become an important added value for the product and the producer.

There are today no real examples of an application of this concept and of these ideas; probably their implementation will require some sort of rethinking of the shoe products as such and of the way the many components of its design concepts are amalgamated to give the products its final shape and nature; and how much each of them weighs in relation to the other.

1. Adidas - MiAdidas [date of survey: November 2006]

CUSTOMIZATION LEVEL	fit and style-aesthetic; feet measured with scanners; best-matched fit approach. Function customization through a simplified gait analysis
ORIGIN OF THE COMPANY	innovative shoe maker; global brand
PRODUCT TYPE	upper class sports shoes
PRODUCT POSITIONING	true mass customizer; affordable luxury
OPERATIONAL APPROACH	click and brick - customized shoes sold in special events in real shops; complementary Website; outsourced manufacturing facilities

Adidas offers, in its MiAdidas program (Fig.4.19), four kinds of shoes (running, soccer, tennis, indoor) with three areas of customization: fit (length and width of each foot); performance (outsole and midsole options and seasonal upper materials); and design (choosing from over 100 colour combinations and embroidered lettering). All of which has to be done in person at select Adidas store locations.

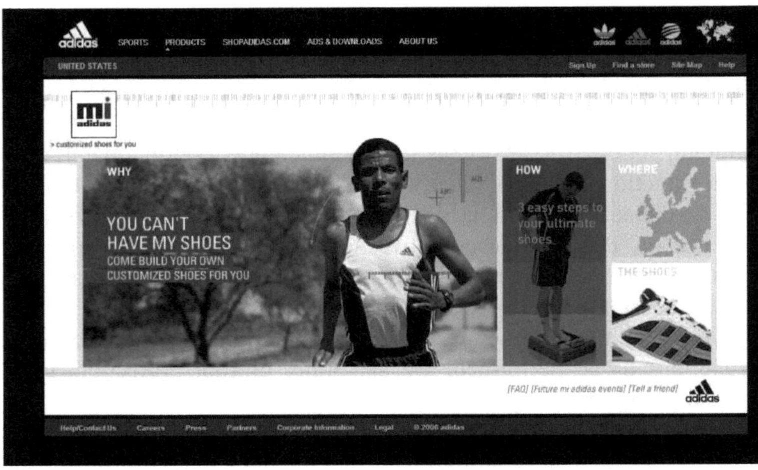

Figure 4.19. The MiAdidas Website

Company profile

Today Adidas groups various brands that cover different aspects of sport goods; in the recent past (1999) Adidas integrated new brands: the Taylor Made Fire Sole clubs and the Salomon in-line skates were successfully launched during the first half of 1999. The original company was registered in 1949 by Mr. Adolf (Adi) Dassler (Das). In 1954 the German soccer team beat Hungary wearing Adidas football boots, made for the first time with removable studs; this event was dubbed the "Miracle of Bern". That moment can be seen as the affirmation of the value of Adidas footwear in soccer.

In the 1980s, after Adi Dassler's death a moment of transition saw his family members carry on the business. The daughter conducted so successfully the company's business in Russia and the Eastern bloc countries that during the cold war the Russian athletes wore Adidas footwear. In the 1990s the management of the company passed to the CEO Robert Louis-Dreyfus; with that step Adidas moved from being a manufacturing and sale-based company to a marketing company that mostly concentrates on product design and development, that controls sales and distribution, but with a totally outsourced manufacturing process.

Passage to Customized Products: When and Why

The introduction of customization started in the year 2000 with the kick off of a Pilot project called MiAdidas. In April 2000 the Adidas board gave a first go-ahead to invest in the shoe customization (MiAdidas) project. But at that time, since the project was a pilot one, the product for testing was only one football boot and more precisely one model in that category: the Predator Precision. The customized product was tested in countries where soccer is an important sport and has a large number of players and supporters, such as Spain, Italy, the UK, Germany, France and Benelux. The launch of such a product was made during sport events where famous soccer players were present and interested in having such a personalized shoe made for them. In 2003 the first retail units were involved, still following the temporary approaches. Then in 2005 permanent units were opened in different shops. Today there are around ten of them around the world including the flagship stores of Paris and San Francisco where the most advanced innovations relating to the sale of customized sports shoes, are shown. This has also involved a restructuring of the supply chain and the creation of a long-term permanent development team dedicated to the follow up of the project.

The added value in MiAdidas is considered to be, for customers, the possibility to perform in shoes produced to their exact personal specifications with regards to fit, performance and design. This was previously limited to professionals such as David Beckham, Zinedine Zidane, Justine Henin-Hardenne or Haile Gebrelassie (some of the most important testimonials of the brand). This new experience is the chance to make technology and freedom of choice available to the common customers. It is a customization experience that provides a chance to create a personalized fit, performance and design for every professional and non-professional athlete. More than just a chance to select colours, MiAdidas is the opportunity to tailor individual shoes to the exact performance specifications of every customer.

Customized Products and Marketing

Today there are four sports shoe types that are offered as MiAdidas:

- running
- tennis
- football
- indoor (handball).

Customers who approach MiAdidas shops are normally very well informed and educated; in any case, posters and information material to present the process and its benefits are available in the shops. Retailers who participate in the program are also very well informed and receive a full package of material and instructions as well as a basic training on to how to handle the process and how to contact the customers, propose the customized shoes and present their benefits. Examples of the components of the shoe and of the materials are available in the shops as a proof of the quality of the products.

MiAdidas sports shoes are sold through a variety of approaches: at specific sports events, at retailers, and at Adidas brand and flagship stores. The company has a multi-approach marketing strategy. Promotional activities include:

- presence at specific sport events (like the Berlin marathon) with direct invitation to a large customer database inviting them to experience the customization process

- partnerships with retailers at whose locations special temporary events are organized
- fixed installations at almost 30 Adidas stores around the world
- fixed installations at a selected and limited number of retailers
- in the future, also through online sales through the Web.

The Customization Process

The customization process of the sport shoe, as performed for instance in Adidas flagship stores, relies on the most modern technologies to support the various steps of the process and to "capture" the interest and the motivation of the customer at all steps. When the customer enters in the shop a futuristic station welcomes him. The customization process is managed by well-trained and knowledgeable shop assistants who lead the client in the process, step by step, explaining all the different stages and storing all customer's data in a database to be recalled later. The assistant uses a tiny palmtop PC that "speaks" with the different components of the MiAdidas sales environment.

The first step is the check-in station where personal data (such as name, address, phone number, age, height, weight, sports activity and e-mail address) are typed out by the customer or shop assistant using a dedicated kiosks in the shop. Then the real customization procedure begins. The main steps are described next, using the example of a football boot.

Step 1: Mi Fit

This is the first step of the process in which foot measurements are taken to identify the best-matching size of the last and data for the functional customization are collected. The fit procedure includes a static and a dynamic phase.

The customer first goes to a static foot measurement system: the WMS system measures the width and length of the right and left foot; the device used is very simple and automatically acquires only two measurements of the right and left foot separately; such measurements are considered adequate for the last fitting process that comes afterwards.

The customer then moves to the dynamic measurement, called the Footscan stage: the customer runs over the Footscan™ mat, and the main features of his or her running are acquired by the electronics of the system. This provides information about the customer's rolling characteristics.

At this stage the real fitting process starts:

- The customer is next asked what kind of stud profile he or she prefers: firm ground (FG) or soft ground (SG). A sample picture is shown. This is basically a choice between replaceable magnesium studs (SG) or a moulded profile (FG). The customer is also asked what kind of field he or she plays on and what kind of cleats are used.
- The customer is then asked what kind of fit he or she likes around his or her toes: short/tight, normal, or loose/long. "Short" implies a direct feeling of the material around the toes, "normal" would be a bit more roomy, and "long" would suggest a lot of space for the toes.
- The width preference around the forefoot comes next: tight, normal or loose. "Tight" really means a snug fit with tight contact around the ball,

"normal" means some contact, and "loose" a really roomy, 'comfortable' fit.
- A question about socks follows; this has an impact on the room in the forefoot area.
- The last question of this phase is about orthotics: does the customer wear orthotics? If he or she does it would be ideal that they be brought in to replace the standard insole during testing.

Static and dynamic measurements and this accurate profiling of the customer provide all the necessary information for an accurate fitting of the chosen football boots. The computer now generates a suggestion based on the data from the foot measurement and the answers of the preference questions: one number for the right, one number (independent) for the left. It is interesting to note that these numbers do not necessarily correspond to the traditional "shoe size"; they rather identify an instance of a size/fit combination and a physical location in the shelves on which the try-on shoes are stored in the shop warehouse.

The shop assistant retrieves from the warehouse the size suggested by the software and brings them to the customer to try on. These try-on boots are considered necessary at this stage, but Adidas aims to eliminate them through the development of appropriate software technologies. The main reasons for getting rid of the test shoes are the costs implied in having a set of try-on boots (something like 100 for each shoe type at each shoe shop) and the error risk they bring with them (it is still a manual process that can induce mistakes). The customer is then able to try some different sizes or width if he or she doesn't feel comfortable with the original choice of the system. The typical selling price of the running shoes is €175.

Step 2: Mi Performance

This is the step in which the "functional" customization takes place; the customer can choose from different studs, sock liners, tongues to have the right shoe for his or her needs; in some models special inserts in the midsole can be used to correct bad running postures and to enhance the performance of the shoe The running test is used in particular to determine the inward/outward tendency of the customer and to then suggest special high rigidity inserts in the sole to correct such tendencies. For running shoes also a summer or a winter mesh can be selected for the upper as well as the type of sole: more resistant or with more cushioning.

Step 3: Mi Design

In this part of the whole procedure the customer can actually take control of the process and design (configure) his or her personal football boots. The customer can choose various colours for the various parts of the upper: he or she can click on the colour schemes on the left and see the immediate result on the right. The software application used at this stage is very similar to the online product configurators used by most of the competitors of Adidas in their Websites. A relatively traditional application but, at least in the version visible in the Adidas flagship stores, with a very innovative user interface; in these shops the user interacts with the computer on a very wide LCD screen by using a finger to point at the screen instead of using a computer mouse. A very impressive and intriguing piece of technology.

After selecting the colour of the studs/ground plate the customer can choose which a word (1 to 8 letters) to have embroidered on the tongue and what colour it should be. He or she can also decide whether to have a playing number and a flag embroidered on the tongue or not. At the end of the configuration process the finished shoe is shown spinning on the screen with the embroidery selection standing on it. As an additional technological "gig" Adidas uses in the flagship store a futuristic device called the "*virtual mirror*": the customer stands in front of what looks, at least partially, like a mirror; in the bottom part, the mirror is replaced by another LCD screen which shows the customer wearing the shoe that has just been configured, with the selections and the colours made combined with his or her real outfit.

At the end of the process, if the customer confirms the selections made, the order is placed and the shoe will arrive in 14 to 16 working days at the retailer. Two certificates are printed out: one for the customer, and one for the retailer. They contain a summary of all the information that was gathered during the MiAdidas experience. It is a proof of the order and has to be signed by both parties to finalize the buying contract.

Orders are collected and validated locally. When this happens at retailers, they go through their ERP systems; at a second level, orders are consolidated centrally at Adidas and processed by their information system. Adidas is planning a complete migration to a dedicated IT procedure for all the steps of the supply chain (including procurement); today some of these steps are still done by hand. The customization process can be summarized as follows:

1. Running scanner, physical measurements, data repository, feet measures/detection of feet problems in running or position of the foot on the ground.
2. Model selection: one of the four models linked to four precise sports and uses: tennis, running, indoor sports, soccer.
3. Aesthetical design of the shoe: colour, design and details of the shoe, finish lining.
4. Payment of the shoe (100%).
5. Order and data are sent to the production factory in China.
6. Customer waits 3 or 4 weeks for the shoes.
7. Delivery of the shoe to the customer at the address specified.

Production System

MiAdidas shoes are produced in China by the same factories that produce the standard models; for example the "customized" Predator football boots are manufactured in the same plant where as the regular boots. Batch and single orders are processed in the same line. The MiAdidas project foresees different steps for the organization of the production side of the business. Initially customized shoes are produced on the same lines as normal shoes, so that personalized and standard shoes are produced together. This was adequate when the level of production was quite low. As a second step Adidas is planning to move the production of the personalized products to dedicated manufacturing lines within the same factories; this is the phase Adidas is tackling now. As an ultimate goal, Adidas is evaluating the possibility of concentrating all MiAdidas production in one, dedicated factory. This will very likely follow the growth in sales of the personalized shoes, which have already reached a few thousands pairs per year.

Customer's Perspective

As it was mentioned at the beginning of this analysis, the main goal of the MiAdidas initiative was that of creating personalized fit, performance and design for everyone's feet. Very much in line with the mass customization paradigm, the purpose was to bring the level of service normally reserved to top soccer players and athletes, also to non-professional sports amateurs, by means of a well-designed pipeline that could keep the selling price of this "service" low enough. Adidas is consumer focused. That means they continuously improve the quality, look, feel and image of their products and their organizational structures to match and exceed consumer expectations and to provide them with the highest value. Adidas is also keen on innovation and design leadership to help athletes of all skill levels achieve peak performance with every product they bring to the market.

Case Assessment

Strengths:

- 100% in line with Adidas vision
- Unique and innovative business model with no real competitor
- Performance image
- Two years experience
- Increases customer loyalty
- Serves vertical credibility

Weaknesses:

- MiAdidas awareness
- Complexity of business model
- Inefficient customer interaction
- Lack of clearly defined responsibilities
- Dependency on retailer commitment
- Leaving information gathered from the customer unused
- Lack of communication and flow of information

Opportunities:

- Using Website as an interactive medium to gather customer data,
- Using the Website to allow re-orders and exchange information
- Add more types and categories
- MiAdidas apparel
- MiAdidas in own stores
- Demand for customized shoes is increasing

Threats:

- Risk of copies
- Complexity of business model is increasing
- Depending on countries' investment

Future Development

In the medium term Adidas plans to extend its leading market position in Europe and Asia, to expand its share of the US footwear market and be the fastest growing major sporting goods supplier in Latin America. A further step in the MiAdidas concept is going online. At the moment the Website is only informative: there is no possibility of buying on line or re-ordering products after a visit and a measurement in shop, not even to configure the shoes online. Adidas is therefore developing the online shop (similar to Nike and the other sports and casual shoe manufacturers) where customers can go to re-order shoes or to order new models after they have been once in the shop. The addition of heritage products is required, to extend the range of customized models, in order to offer its clients a wider range of products to choose from.

1. **Fila - Adatto** [date of survey: March 2007]

The process of documenting how a new business model or a manufacturing paradigm is penetrating an industrial sector tends to be a never-ending one; rumours of newcomers introducing new lines of customized products, or of projects on the verge of being started by footwear companies all around the world, have circulated during the entire preparation time of this book. We have decided to mention one of them here for two reasons: it refers to a producer of sports shoes, relatively close in terms of product typology to Adidas, and it deals with various levels of customization, including functional.

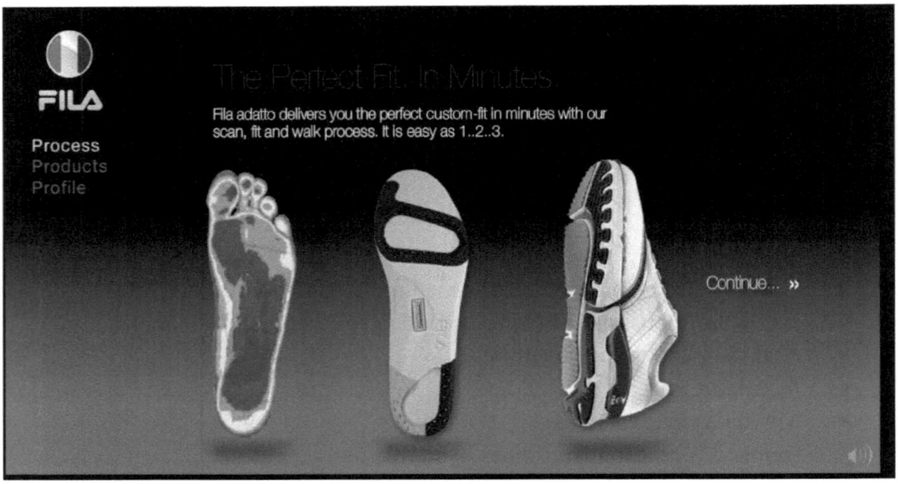

Figure 4.20. The Fila Adatto presentation page

The company is a famous Italian brand, Fila, known mainly for tennis shoes but, also, for sports and casual footwear. The company is just coming out of a negative economic phase and has been investing significantly over the last ten years in products and marketing. An announcement was made recently about the project Fila is launching in the field of customized sport shoes (Fig.4.20).

The product line will be called Adatto and, according to the claims of the company that have appeared in the press in the last few months, customers will be able to buy them in dedicated shops. A foot scanner will be used to acquire foot data and at the end of the measurement and configuration process, according to the claims, shoe components such as the foot bed and the upper will be immediately "manufactured" in the shop.

We thought this news was worth including especially for that aspect; shoes will not only be configured and customized in the shop but also manufactured there; a far deeper and more complete implementation of the mass customization paradigm. We will have to see how far the promise will be kept when the Adatto shoes will go on sale (March 2007, with a press campaign, according to the declarations, which will be worth €3.5 million), but whatever the real implementation will be, the Fila-Adatto project, on which the company declared it has invested €4.5 million in research and development, is certainly one of the most important and impressive industrial projects in the field of footwear mass customization.

5

From Footwear to Other Industries

5.1 Peculiarity of Mass Customization and Footwear

In the previous chapters we have thoroughly presented all the relevant concepts related to mass customization and to its implementation in the footwear world. Prior to the final Chapter of the book, in which we will try to draw some conclusions on the impact of MC in shoe companies, we want to address some additional matters that we think can complement the ideas and the concepts that we have developed so far and lay the ground for the final conclusions.

As we have mentioned many times, mass customization can be applied to almost any product in a variety of ways and with a variety of business models and, in particular, with a variety of customization levels. The footwear sector have some peculiar aspects that make customization quite particular and, in many respects, different from other industries. The product must fit two parts of the human body that are in a certain sense very similar in billions of cases but they are also always different! Furthermore, the mass customization solution for footwear is complicated by the fact that the product, the shoe, is also different by function, style, price, quality, *etc.* Furthermore, health problems and concern may influence even more the customization in the footwear sector. Even if we take into consideration that roughly 30% of the population in western countries is really interested in customized shoes (as was demonstrated by the study made in the EUROShoE project), this still represents hundreds of millions of shoes per year as a potential market. It is difficult to find any other product with such a huge potential market and with such complexity in the value chain from design to manufacturing and delivery where the customer is at the centre of the whole process.

Do we all intend "mass customization" in the same way? Unfortunately not. Mass customization has become almost a buzz word and therefore there is the tendency to declare some business models as "mass customized" when instead they are not. Let's take the example of Dell Computers. It is very often cited as the example of customized computers for single individuals. You can go onto the Dell Website and build a computer that has only what you want - nothing more, nothing less. Want more RAM? No problem, add more RAM. Need a DVD instead of a CD-ROM? Not a worry. Got to have that extra disk space for all those MP3s? Not a concern, just choose it. With only a few clicks of the mouse, you get the exact computer you want, not what just happened to be in stock. But, how can Dell have so many versions of one computer? Simple: they don't. This is all possible because you are buying a computer that does not yet exist. Rather than sitting on a shelf, waiting for a customer to take it home, the computer you buy at Dell only exists as a bunch of parts that someone at Dell is going to build, sell and ship to you in less than 24 hours.

Customization, or mass customization according to many authors, is a way of building and selling products such that the product features are broken down and offered to the consumer as choices. For example, if Dell has 20 or so product features that make up every computer (RAM, disk space, processor speed, modem, operating system, *etc.*), the customer can pick and choose between all 20 of them, resulting in a computer that is customized to that particular customer's needs. However, this process is called *assemble-to-order* not mass customization.

What is the difference? The difference, for example, with the footwear mass customization is substantial in one and only one particular aspect: while the Dell computer exists already in all its various main sub-assemblies and it will be mainly assembled in its final composition as soon as the order from the client arrives at Dell, a pair of shoes that are truly custom fit for the customer do not exist at all. All (or most) of its components must still be manufactured and assembled. So the same process used by Dell cannot be applied, as such, to customized shoes. Even if the shoes are made in the best fit way, the process is different. The selection of the best fit shoes requires a measurement of the feet of the customer that a PC does not need; it then implies a level of interaction with the end consumer and a thorough understanding of his or her physical peculiarities and individual desires that is hard to find in other sector (with the exception of apparel).

Of course some footwear are also assembled-to-order, but they are not made on a specific form of the customer feet. There is also a confusion between *made-to-order* and customized product. For example, the one-of-a-kind engineering products are individually made, fitting some given specifications but they can, in principle, be repeated many times for different customers. Instead, a pair of shoes made for one customer cannot perfectly fit the feet of another customer. However, we may say that mass customization is the synergetic merger of known old approaches [1]:

- *modern manufacturing* as theoretic background
- *modularization* of product architectures and use of product platforms
- *limitation of customization possibilities* reflecting demand for a stable processes
- *made-to-order approach* - split of fulfillment systems into standardized and customer specific parts (decoupling or postponement point)

- *e-Business principles* - use of dedicated information systems for configuration, manufacturing planning, order tracking, and relationship management

While some customizable products seem interesting and useful, and others do not, the question is raised of what really can be mass customized? Where will the limits of technology and customer demand stop? You can already order a house from a series of menus and toolbars displaying choices; will the rest of the real estate market follow suit? Soon you will be able to get a customized Italian motorcycle, but will we one day order our own cars from the Internet with the headlights, seats and body frame that we chose? Will buying a washing machine be similar to buying a Dell computer? While many would argue that the answer is "probably", and that the sites for customized products (see also [1]) available today are only the tip of the iceberg, one thing remains true: no matter how advanced the technologies and possibilities are, there will always be some products that are simply made one way in one size for all interested customers.

This is true for several reasons, including the one that people love to shop in stores, and always have and always will. Some products (such as a computer) will have so many different permutations that it will be impossible for a company to offer them all in a retail setting. Another strike against mass customization is the fact that some people don't want choices. Customizing products will not only put more demand on the consumer to take part in the process, it may even be off-putting to some. Designers as well are likely to take issue with demands for their products to go on the market untouched and unscathed. Yet, despite the obstacles, it is clear that mass customization and the products that can be customized will grow over the years. What does that mean for consumers? More choices, more "interactive shopping" and more personalized products.

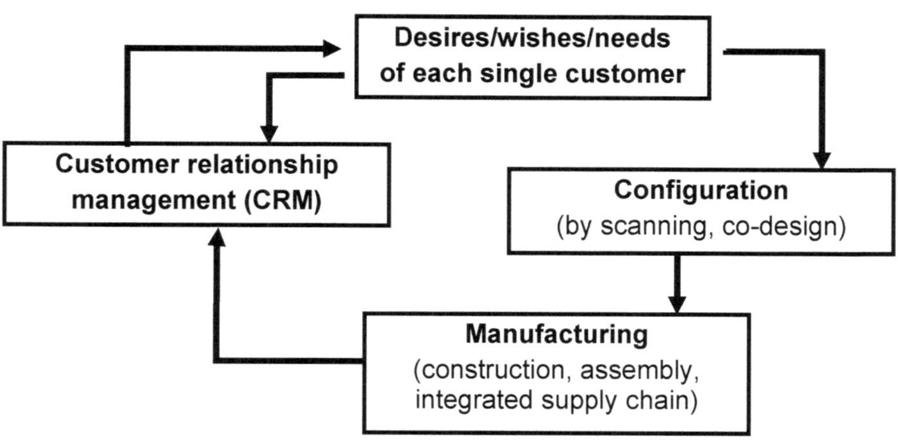

Figure 5.1. Mass customization for footwear

However, the difference with mass customized footwear appears to be clear: it combines all the good aspects of true customization. For example, the customer will still go to the shop to choose a pair of shoes, and the designer's style will still be respected, but the consumer will feel more secure having the shop assistant help to measure precisely his or her feet to adapt the selected shoe style to his or her unique feet form. Therefore, the mass customization paradigm for the footwear industry can be described as in Fig.5.1.

As can be seen from the diagram, one of the fundamental characteristics of mass customization for footwear is the configuration of the shoe by scanning the feet of the customer. The configuration of the product in other cases or sectors is much simpler or it must be standardized in order to reduce complexity (PC, car, watches, *etc.*). One sector that is very close to footwear is textiles. Also in this case the body or part of the body of the customer have to be scanned. However, the nature of the product (shirt, jacket, trousers, *etc.*) is such that the flexibility of the material allows much more tolerance than in footwear.

Hence MC fits to footwear in a peculiar way that needs to be clearly understood when the general specifications of mass customization are to be applied to shoe making. Some of these specifications have already been mentioned, but we want to highlight here what we think are the most relevant distinguishing aspects that make MC in footwear quite different to other implementations.

5.1.1 The "Human Touch"

The implementation of mass customization in footwear is peculiar, first of all, for the fact an interaction with the human body (or at least with a part of it) of the consumers is needed; although, thinking back to the definitions given in this book, this is not true when only style customization is offered, as market and consumer surveys have shown that MC gets really interesting when fit customization is available. And in such cases a scanning of the foot is required, which implies a physical shop, some kind of assistance from the sales personnel, and a good level of attention to the identification of the needs (objective) and desires (subjective) of the consumers. All this calls for the "human touch" in managing the relationship with consumers that makes footwear MC peculiar. The psychological implications, from the side of consumers, of lending a part of their body for some kind of "data processing" (a consideration that is certainly true when we think of footwear but which becomes very true when garments are considered) are not to be underestimated (think for instance of all the privacy implications related to handling biometric data such as those produced by a foot scanning) and they are a further factor of differentiation in the application of MC to footwear.

5.1.2 Dedicated Manufacturing

As it has been mentioned, shortcuts in manufacturing (like assembling to order) are not so easily possible in the case of footwear; even when the simpler Best-matched fit approach is adopted (see Chapter 3), there is a great number of components of the shoe that cannot be pre-produced and kept in stock - more than 50% of the most relevant components of the shoes need to be manufactured on order. This is not only driven by the dimensional variability related to the size of the shoe to be made, but also to the unpredictability of the choices the consumers will make in terms of colours, materials and components of the shoe during the configuration process. The manufacturing of customized shoes is very little "assemble to order" and very much "made to order". This introduces another tremendous complexity factor: traditional manufacturing units are equipped to handle long runs and large batches, and are not very well suited to handle individual orders. Hence, managing traditional and customized production in the same production line tends to be a desperately difficult exercise. The best solution would be to rely on dedicated facilities, specifically designed for MC and dedicated solely to this business.

5.1.3 Fast Changing Styles

Dell computers, cars, or other items that are being offered as customizable products, are relatively stable in their design; new or better performing components (a faster processor or a larger disk) can appear, but the very "architecture" of the product does not change very often and the life of the modular products is counted in months if not years. A shoe is a fashion accessory and thus follows the fashion trends; it changes from season to season, it evolves and it adapts its shape, its construction and its appearance to the new trends. This happens twice a year in the easiest situations, four or more times a year when trendy and fashionable models are offered. This means that not only modularity aspects are to be considered for the customized shoes, but also how these modularity aspects are affected (or may affect) the changes in style dictated by the fashion trends.

5.1.4 The Speed of Complexity

The above topics introduce an additional factor of peculiarity in mass customization applied to footwear: a shoe may not be in absolute terms the most complicated product (in terms of number of components and of their variability) - a car or an aircraft can definitely be much more complicated than a shoe, with bills of material that count millions of items rather than a few hundreds - but these millions don't change very fast, since new models are introduced very slowly and in time frames of years. Shoe collections, as we said, change at least every six months and the whole effort of designing them for MC, of handling their manufacturing constraints, of planning all the necessary processing steps, is repeated with the same frequency. It is the speed with which complexity evolves that makes the difference, rather than the complexity in absolute terms. This again makes customization of footwear so peculiar.

If these are the peculiarity aspects of the implementation of mass customization in footwear, we may wonder if there are lessons that the footwear world can teach other sectors in their path towards the new business model.

5.2 What Mass Customization and Footwear can Teach Other Industries

Certainly, if there are peculiarities, there are also many factors that can be regarded as common to other sectors, such as:

- new differentiation possibilities
- value competition instead of price
- reduced planning risk, increased flexibility
- new cost structures
- reduction of inventory, fashion risk
- new dimensions of shopping experience
- customer knowledge
- lead user information
- access to "sticky information"
- higher market research efficiency.

These topics, if well exploited, can turn into highly valuable competitive assets both for shoe making and in general. During the EUROShoE project a marketing survey was performed and the results may be of interest also to other products and sectors. Some of the most interesting outcomes of that research are listed here.

- For the consumer there are different degrees of acceptance but no difference between manufacturer and retailer.
- Scepticism and refusal dominate the attitude *only of the interview partners who never considered customization of shoes.*
- *Fit and comfort* are the most important criteria for the buying decision.
- Customers want a *comfortable fitting shoe* with a strong *fashion content.*
- *Comfort shoes, modern classic men shoes* and *sport shoes* are interesting for customization, before any other model.
- By far the most important customization parameter is *"fit"* and "comfort"; referring to *design: colour, shape, material, heel shape and appearance* are the most relevant ones.
- *Comfort oriented customers, individualists, fashion oriented women, young people* (sport) *and "classic business men"* (executive position) are interesting target groups.
- A *premium* for customized shoes of *20-40 %* is acceptable.
- There are different degrees of acceptance, but the mentioned advantages and disadvantages regarding customization are very similar in the four target countries (Italy, Spain, UK, Germany) analyzed in the EUROShoE study.
- Customers need to be guided carefully to the new concept – most of them are *open minded* but cannot imagine what is going to happen and how when they are supposed to purchase a customized pair of shoes.

- *Perfect fit* is what most people want; design customization is especially rejected in southern Europe where a higher attention to fashion and its trends is present.
- The most frequent consumer desire is: "To make the *fashionable* shoe more *comfortable*!"
- Customers need competent *advice* and a *variety of customizable models*; the Internet is only considered an add-on sales channel.
- Standard trial models must be available in the shop; *touch and feel are important*, also indicating the need of *retail-based concepts*.
- *Delivery time* of one week, or a maximum of two weeks, is acceptable.
- Premium price is accepted if the added value of the customized shoe *is fully perceived.*
- *The majority of European consumers are interested in customized shoes (more than 65% of women and 55% of men).*
- 29% of women and 46% of men interested in *customised shoes* would buy shoes *on the Internet.*
- Most customers would accept a customization time (configuration) of *20-30 minutes.*
- For formal and casual shoes, most men and women accept a *premium of 10%-30%* above the standard price with a slightly higher tendency with formal shoes.
- *Fit, comfort and design* are the most important parameters to customize
- The interest for customized shoes depends on the *encountered difficulties* when looking for standard shoes.

By reviewing this list of topics, we can clearly see that many of the statements that they contain have a very general nature and can be easily applied to other sectors; in particular to those sectors, like the apparel one, which share with shoes and shoemaking aspects such as style and fashion dependency, their influence on consumers' comfort and well-being (both in physical and psychological terms), the complexity in their design and manufacturing processes. The lessons learnt in investigating MC and footwear can be, in this case, almost immediately transferred to other similar sectors.

But there are other interesting factors that can be analyzed in relation to the acceptance by consumers of the concept of customized shoes; the understanding of these factors is very important for the successful implementation of an MC project in the footwear field as well as in other fields. A study of this nature was conducted as a complementary activity of the EUROShoE project; the work [29] was aimed at analysing the cultural gap that apparently still exist in the acceptance of mass customization (and in particular of mass customized shoes) by consumers.

The outcomes of this study are very interesting and despite their origin (the EUROShoE project and mass customized shoes) they have a general validity that makes them very valuable also for other products and other industries. The survey was a good outgrowth of the EUROShoE project which was concluded in September 2004 within a validation campaign; in this phase of the project a few hundred of persons living in the neighbourhood of the location of the Pilot Plant (see Chapter 3) were invited to test the collection of customized shoes that were developed and produced during the project.

Telephone interviews with 25 women and 25 men who experienced the customized shoes were carried out, using open-ended questions to establish a rapport with the interviewees, and to easily capture suggestions. These questions were aimed at investigating how these "sample consumers" perceived the concept of customization, how much appealed (or scared) they were by the new approach, and what was their general acceptance of the personalized products. In this section of the book we will present some of the most relevant outcomes of the survey, listing the questions that the interviewees were asked and the nature of their answers.

Do you know the meaning of "customized"?

All the interviewees answered positively: "Yes, I know the meaning of that word, I learnt it when I was invited to participate in the test"; another said: "it means to have a pair of shoes produced using the computer." A retired man added: "Yes, I know the term. When I was young I was a football player and I remember the craftsman measuring my feet with a tape to make me a pair of shoes." But the more complete definition of customized shoes was given as: "Yes, it means that the shoes have to fit on my feet and not the opposite, as normally happens …"

Is there any difference in the production process of a customized shoe compared to a mass produced one?

Almost all the interviewees thought that some differences should exist in the production phase, even if they did't really know which one. Actually some of the interviewees managed to tell what they thought was the difference: some of them said that customized shoes are produced more carefully, which is the reason they are more comfortable. Others focussed their attention on the scanning phase, and others guessed that some of the machines are different. The simple perception is that the customized shoes should have some links with craft-made products: "the shoe is sewed with a sewing machine, not with a robot", or "there are some operations that are certainly handmade."

What kind of problem can a customized shoe solve?

"It can solve specific needs, from non-normal feet requirements or some malformations …", answered a young interviewee. Actually the opinions were split between the comfort issue, "if the proportions of the shoe are correct, balancing length and width, the shoe is always perfect", and the solution to some minor problems of the extremities: "a customized shoe can help the setting-out of the walk, avoid bad posture of the feet and minor foot pathologies." Even if the concept of "comfort" is subjective, because each person has their own meaning for this experience, all of the interviewees fixed their attention on better quality of walking; only one of them highlighted an aesthetic factor: "… moreover, the possibility to choose the aesthetic aspect you prefer."

What kind of problem can't a customized shoe solve?

Many people declared they did not know the specific problems that remain unsolved using customized shoes. "Customized shoes can overcome any problem, because of the materials: the leather is so soft nowadays that, if the shoe is well made, all the problems can be solved." And, if a subject in the previous question underlined, as a positive aspect, the possibility to choose aesthetics, here another, quite in opposition with that concept, indicated that customized shoes don't go together with beauty and elegance.

Normally you should pay for your shoes in advance, before having them in your hands; what kind of warrantee should you like to have?

Trust in the producer implies that the users have already tested the product and its quality, as the interviewees did during the EUROShoE test campaign. There are three main discriminating factors: *the price* - people show anxiety concerning the higher price of the product ("it depends a lot from the relation between quality and price"); *the quality* - customized shoes have a specific characteristic: a better quality as the previous question has already showed; *time to delivery* - some interviewees raised some time concerns ("too much time passes between the order and the delivery"). Some wanted to pay at the time of the delivery. "In a pair of shoe I look at the aesthetics, the leather and materials: I don't want to pay for my shoes if I don't see them!"

When you came to the "customized shoe shop", did you miss the presence of a shop clerk?

The majority of the negative answers completed their statements with: "I hate clerks", or "I don't need clerks, I can decide on my own." But one of the persons, mentioned something different: "I don't miss the clerk. The PC suggested a specific pair of shoes, that is comfortable, but if I could choose, I would have picked up another model." Among those who answered positively, one of the interviewees mentioned: "I missed very much a clerk! ... to have a suggestion ... The human contact is something basic!"

If one day you were able to buy your shoes using a PC, via the Internet, would you like it?

The negative answers to this questions were usually motivated by the need to look at a real collection, and to touch the product at least. "I miss the human contact. I don't want to have a relation with a machine. But this is probably a generational matter", some of the older interviewees stated, but even a young person explains: "I don't buy anything on the net. Even for books: I prefer to go to the bookshop and browse a book." Just two of them admitted their incompetence in using the Internet as the reason for their negative answer. Many people in the positive group added: "I had some doubts at the beginning, but, if I can buy a first pair of shoes in a shop, then I can order the same using the PC with no problem!", and another one reinforces: "I can use Internet to re-buy a well-known model." A secondary clause for some positive answers is the trust in the producer: "I would use the PC to buy from a reliable firm."

How can the customized shoe concept be publicized? Which channel can be used?

The answers to this question were quite varied. The majority of interviewees indicated newspapers with TV coming second: "it depends on the penetration you're looking at. In our society, today, TV is the only mass media that arrives to everybody." But someone thought that was important to reach a special target, not the mass, for this reason: "you can use specialized press, to contact a selected target." Interesting is the suggestion to organize "workshops and demonstrations. Maybe in shopping centres where the users, even if they don't buy anything, can watch and ask for information."

How can an educational action be made, explaining the importance of a customized shoe?

In this area medical doctors appear as well as locations devoted to teaching activity, where "the word *health* is very common: it's as a magic word and it's perfect also for advertisements", and also the workshops "should be held by a specialist, maybe a podiatrist", said someone else. An interesting possibility was mentioned by one of the interviewees: "You can use fairs as occasions for making the retailers aware and not only the final users. You can also use associations to inform the mothers and not only the children. Finally, medical congresses." Only one person indicated that information on the effect of a good walk is not so important: "Since the people who needs customized shoes are not so many, because those who have normal feet don't need it, information isn't necessary at all!"

The computer decided the model of shoes for you; no try-on tests were made. Did you miss that step?

Those who said they missed the physical fitting of the shoe would like to feel the comfort; some of the interviewees said: "I would have had a kind of fitting, as I can do with a new dress sewed by my tailor." Even most of the people that asserted they didn't miss trying on the shoes, added that they would have liked at least to see a real sample, having it in their hands to verify the colour variants of the leather and the quality of the construction of the shoe. "No, I didn't miss the fitting moment, but I would have wanted at least to see and touch the collection" - this means that a sensual feeling with the product is essential. Among those who declared they did not miss the physical test, one answer was particularly worth of notice: "No, I didn't miss it. On the contrary, I liked very much that the computer had chosen in my place. I'm not able to decide on my own, and I trusted the system."

In which moment of the production process do you think it's important to give your opinion as a final user?

An early participation in the process was the most typical answer, yet other interviewees gave different points of view: "I would participate in the entire process" and "I would like be present at the beginning for the aesthetic configuration, and at the end when I fit my shoes and I can give my opinion on the comfort." Moreover, there were some who would have liked to have "a kind of intermediate fitting, in the middle of the production process", probably biased by previous experiences with the customized dress of a tailor. Some of the interviewees declared that "the production phase is something far from their experience", for this reason they were not able to answer; but there were some others who stated: "if I could, I would intervene also before the design phase. I would like to assemble my shoes starting from different pieces and elements, like a puzzle."

Do you trust the accuracy of the price indicated by the producer?

Typical comments were of the kind: "I have some problems: the price should be significantly different from the market, and I couldn't compare it." This appeared to be the key question: the customized product is something relatively new for the market, and the price range is an unknown factor. One interviewee added: "I can evaluate the price with the product in my hands. If I'm satisfied with the result, I can even accept a higher price!"; "I want to be safeguarded! The producer should be certified. And I must have the possibility of sending back the product that I don't like!" One of the interviewees proposed a solution: "The first time the producer should give me a special bonus for paying with a discount, if I'm happy with the product, then, I can pay even more!" But eventually the matter is once again trust in the producer: "This is the reason I prefer to shop in shops that I know well: I have confidence in the owner, I know him, and he knows me, I experienced the materials and the quality of the product. I've always bought there."

Which segment of population need more customized shoes?

Many people declared that children are the principal group candidate to customized shoes because they "have feet that are growing and their walking need to be appropriately helped", but some others thought: "I think customized shoes are important for all, men, women, but adults, not children because they grow up too fast and the shoes should be changed very often … customized shoes would turn out to be too expensive." Some of the most educated interviewees said: "Cross-sectional, but middle class, upper-middle class will buy customized shoes, for the higher price, the upper class goes directly to the cobblers."

Where do you normally buy your shoes?

A peculiarity of the area where the interviewees are based, is the presence of shoe shops and outlets of Italian shoe makers. Those who choose the shop, do that because they like the human contact, they "love to be suggested and counselled", and "don't like the cold atmosphere of a self-service."

How long does it normally take you to buy your shoes?

The process of choosing a pair of shoes could be very long, and many of the interviewees declared that they normally spend many days window shopping before choosing and trying on the target pair. There seem to be no such a difference in the buying habits between men and women. But some of the interviewees also confessed: "I only need a few minutes because my wife goes and looks for the right shoes for me; at the end I simply go and buy them".

Do you think it will take more or less time to buy a pair of customized shoes?

The discriminating factor in time related issues is the absence of the fitting phase: "after the scanning, the PC helps me in selecting my footwear" so it takes less time than a normal shopping moment, even if "explaining my feet problems takes more time" or "I can choose the model in a kind of catalogue, and if it's a big collection I need time to choose it." Some of the interviewees were more hesitant: "It depends: if I personalize too many aspects of the design of my shoe, this is going to take me more time. If I choose between some of the predefined models, it takes less time."

We believe that the kind of indications that emerge from this survey, based on the opinions expressed by consumers who had a chance of experiencing the mass customization process, are very interesting; they contain important guidelines for companies who want to move along this path. Although some of the aspects that were touched are very much related to shoes and shoemaking, most of them have a general value and contain important indications for virtually any kind of customization project that aims at a direct involvement of consumers. This is once more an experience that footwear can share and transfer to other sectors.

6

Conclusion: A Few Questions to be Answered

6.1 Has Footwear Mass Customization Arrived?

As we have extensively discussed in the previous chapters, a relatively large number of big brands, start-ups and outsiders have already discovered the potential of footwear mass customization and they are taking advantage of it, learning day by day from their customers. Is this enough to state that mass mustomization has arrived?

From the indications that we gathered there are two evident trends. One puts more emphasis on the customization side of mass customization (MC) and it finds its followers among new or traditional suppliers of bespoken shoes (typically for men) who want to modernize their handmade approach to the consumers, with all the most modern equipments and methodologies that computers and information and communication technologies (ICT) are making available to them. Their products will then be "affordable luxury", as we have classified them in Chapter 4, bringing down to a wider segment of consumers the "top class" prices that traditional purchasers of bespoken shoes are ready to pay. Not really masses though.

The second implementation of mass customization looks more at the masses of consumers, is more adequate for traditional shoe producers, relies on their classical and available manufacturing facilities and really aims at capturing a much wider range of consumers, who never experienced customization, with a personalized product at a price that is only a little more than the one they normally pay. If we had to state the current situation of the penetration of the new business model into the footwear world, we could highlight three well-defined trends:

1. **An inward trend** - by outsiders or by entrepreneurs somehow already active in shoemaking but with a dimension that is far from being "industrial", to enter shoemaking from the point of view of customization. These types of "market players" normally bring new and fresh ideas, are

not biased by traditions or by the constraints (mental more than anything else) that the classical approach to shoe manufacturing brings with it; and tackle the problem of setting up and managing the complex operations of a mass customizing company with an open mind and with a clear "consumer centric" goal they want to achieve.

This group of mass customizers is not normally bothered with controlling, at least in the first implementation phases of their project, the entire design and manufacturing pipeline; they invariably concentrate their interests and their investments: products and the consumer approach, both for the sales and post sales moments. Hence their customization approach is focused on the choice of the appropriate models to be offered, on the selection of the most valuable constructions and of the right materials; this is even more true when these companies (outsiders) are not traditionally coming from shoe making.

Front-end technologies are the next area of big investments in the MC projects of these outsiders; the examples show that in these cases entrepreneurs are much keener to set up a nice and elegant sale point, to install foot scanners and product configurators, to develop efficient ICT platforms, than to put money into manufacturing facilities. A captive manufacturing capability is not considered to be an asset, at least not an immediate one. Yet there is a danger in exaggerating this approach; underestimating the importance to rely on a well organized and comprehensive production facility and becoming too much dependent (in particular when this reflects on aspects such as customers' satisfaction) on the "humours" of the manufacturing suppliers, soon turns out to be the major bottleneck in migrating from the exploratory phases of the project to the full exploitation ones.

2. **An outward trend** - by traditional shoemakers who are leaving (at least partially) the known world of their standard manufacturing and classical sales approach, to the new world of personalized products. It must be said that there are no evident indications that this is happening in the form of a total shift from mass production to mass customization. It rather manifests as a slow, progressive, partial move towards the new paradigm. Shoe companies are (and they more and more will be in future) in the situation of having an overproduction capacity and so they potentially could dedicate this spare capacity to "exploring" more innovative approaches to their traditional and captive markets; on the other hand, companies are desperately seeking for differentiation factors, for added values to characterize their products with, in order to regain at least part of that competitiveness they have lost in favour of other producers. What is happening is that today a handful of companies, with perhaps more to follow, are taking advantage of their presently underutilized manufacturing facilities to transform them into units dedicated to producing personalized footwear; they are of course also investing in the front-end processes, investing in hardware and software to open sales points where their clients could go to purchase the personalized shoes. The higher the control they have on the retail side of the business (for

instance when they own one or more shops where their brand is sold) the easier it is for them to consider the new approach and the higher their determination in pursuing the business idea.

3. **A transversal trend** – this takes place within the footwear sector, along its pipeline. We refer with this comments to what is happening on the retail side in relation to mass customization. We have already made, when we discussed the general motivations for mass customization, some statements on the role of retailers in a mass customization scenario and on the potential that this paradigm can have for them; we want to consider this point further.

   Shoe retailers[37] have been absent from most of the considerations and examples we have presented so far; but something is happening in their part of the business too. A superficial interpretation of the MC paradigm, with the central role of the consumers and the regained direct relationship between them and the producers, would seem to indicate a possible exclusion of shoe retailers from the MC business. In fact, this is not the case; detailed value generation analysis have been done that show how both producers and retailers can take advantage of the redistribution of the business margins that MC produces. We have a feeling that the initial diffidence with which retailers considered the idea of mass customization is now almost completely gone.

   What we have noticed is, on the contrary, a growing interest of some of them to renovate their business with the new ideas and the more modern approach to the consumer that mass customization implies. So there are indications that both retail chains and individual retailers are developing a greater interest for MC; they are thinking of installing foot scanners and product configurators in their shops, and are talking to their shoe suppliers with the aim of setting up adequate supply chains to sustain the new business. They are, on this respect, in an ideal position, being already in daily and constant contacts with their clients; so it is much easier for them to convey the message of customization, to gather the feedback of their best and most faithful clients, and assess the acceptance of the new idea. We can see this starting and we believe it will happen more and more in the future.

   There are other signals that a potential exists; some technology providers (see for instance the solutions developed by Formalogix, although the Slovenian UCS is also moving in the same direction) are offering solutions that are aimed primarily at the retail side of the business. Perhaps in some cases their approach is not exactly centred on the personalization issue, but rather in offering shoe selection or shoe location tools; but they share with mass customization the idea of

[37] We are conscious of the generality of this term and of the fact that retail approaches and retail networks can have very different forms and organization, in terms of structuring and size of their members and characteristics of the shops; and this varies a lot from country to country. But we believe that the comments we make apply in general, regardless of the specificities of the particular country or network.

consumer centricity and the innovative approach to shoe selection that is typical of the true mass customization implementation.

Hence, has shoe mass customization arrived? A honest answer would be no, or, if we want to be more optimistic, not yet. The big move still has to take place; although the interest is growing and the number of companies considering MC as a possible option for the prosperity of their business is constantly increasing, MC is not yet perceived as a "normal" way of dealing with the footwear business and as a source of secure revenues for the companies.

6.2 Is Footwear Mass Customization a Myth?

The second question we want to address in this concluding chapter is whether mass customization, in relation to its achievements and to its general level of acceptance among shoemakers, can still be considered a myth. In order to proceed with these considerations, we think it is worth defining the term "myth" more precisely:

> *In the academic fields, a myth (from the Greek "mythos") is a sacred story concerning the origins of the world or how the world and the creatures in it came to have their present form*[38]

and also:

> *A myth, in popular use, is something that is widely believed but false.*[39]

Rather than the academic and canonical definition of myth that relates it to the *"origins of the world*", which is not certainly our case, we would rather adopt the popular definition that has associated the word to the wide belief of a thing that is not true. So is it the case that the benefits of shoe mass customization are not true? And, furthermore, is it also untrue that shoe companies are interested in it? These two questions require separate considerations, although the answers will differ that much.

Let's start from the point of the benefits: we are convinced that market benefits are evident and that economic return can be proved and documented. The indirect demonstration of the correctness of this statement is the fact that the number of shoe companies that are implementing the new business model is constantly, albeit slowly, growing; they are proving that the advantages of MC are not a "myth", paving the way for the "masses" of companies that still have to move. The fact that the number of newborn companies outweighs the failures, adds to the considerations that the economic fundamentals of mass customization applied to footwear work.

---

[38] Source: Wikipedia.org. Wikipedia is possibly less accurate than the Oxford Dictionary but certainly more widely and popularly accepted.
[39] Same source.

Now coming to the interest of shoe companies in mass customization, we can make the following comments: only a few years ago, namely before the EUROShoE project started, if you asked shoe makers whether they believed in the possibility of producing any sort of personalized shoe at costs comparable to those of the standard ones, most of them would probably have looked at you with a high level of skepticism. If you asked today, although skepticism has not entirely disappeared, but skeptics are an ever-reducing minority. The majority of the companies we had a chance of working with or had the opportunity to discuss things with, now believe that what they thought was impossible, is in fact achievable.

We think that two factors have contributed to this change of attitude. First technology has progressed. Some of the EUROShoE partners as well as other companies have in the last few years developed and released to the market families of machines, hardware equipment and software that have enabled the achievement of a level of efficiency that, on the other hand, allows to keep control on costs and to maintain them at reasonably competitive level. Second, and more importantly the level of information of the companies and their "culture" on the potential, the benefits and the complications of footwear mass customization, have definitely increased.

The companies that have initiated successful mass customization projects, are those that wanted to learn about it; that have dedicated time and resources to investigating its application to their specific business sectors; that have contacted their customers to learn how much they would be appealed by the idea of having personalized shoes; and, eventually, that were convinced that the necessary technologies were available.

This is the point we want to make in concluding this section: the adoption of the mass customization paradigm in footwear is possible; it is possible with a variety of different approaches, with increasing levels of satisfaction for the consumers, each one of them implying higher or lower levels of complexities in the processes that shoe makers have to consider to sustain this new activity. By carefully choosing the "entry point" for its own mass customization implementation, a shoe company can "customize" its specific customization solution, deciding on which ones of these process it should focus, knowing which degree of complexity it will have to handle, and being fully aware of the benefits and of the final returns on investments of the whole operation. And, last but not least, knowing that the most relevant enabling technologies are there and that they allow a level of performance that, although yet improvable, is high enough to let them venture in mass customization.

6.3 Can Mass Customization Help Sustainable Development and How?

At present, it is no longer possible to ignore the impact that industrial products have on our ecological system. The shoe industry has a long tradition in trying to find solutions to hazardous and environmental damaging processes (like, in tanning, the "wet blue" using highly dangerous chrome) and materials. According to the authors of one the best books [30] about sustainable development:

> ... *another object of modern manufacture: the shoe. Michael joked that his guests were wearing "hazardous waste" on their feet, waste that was abrading as they sound on the rough surface of the floor, creating dust that people could inhale. He told how he had visited the largest chromium extraction factory in Europe – chromium is a heavy metal used in large scale leather tanning processes – and noticed that only older men were working there, all of them in gas masks. The supervisor had explained that it took on average about twenty years for workers to develop cancer from chromium exposure, so the company had made the decision to allow only workers older than fifty to work with this dangerous substance ....*[40]

And then the authors go on with other considerations about the design of shoes. The term "sustainable development" emerged almost 20 years ago. The United Nations World Commission on Environment and Development coined the most widely used definition: "development which meets the needs of the present without endangering the ability of future generations to meet their own needs" [33]. The term refers to achieving economic and social development in ways that do not exhaust natural resources. Even though the concept of sustainable development should take in consideration the interaction and integration of environmental, social and economical elements in the global, national, regional or local contexts, the term is usually mainly related to the environment, such as lack of energy, climate change, water scarcity, waste management, *etc.*; as a result, less attention is paid to social and economical aspects to foster sustainable development, specially the importance of businesses' sustainability. In these regards, Professor Ashford from MIT [32] argues the following:

> *It is quite a difference whether you look at sustainable development as just an environmental issue, or alternatively as a multidimensional challenge in the three dimensions: economic, environmental, and social.*

He argues that competitiveness, environment and employment are the operationally important dimensions of sustainability – and these three dimensions together drive sustainable development along different pathways and go to different places than environmentally driven concerns alone, which may otherwise require tradeoffs, for example, between environmental improvements and jobs.

Therefore any "shoe related" industry, organization, institution or individual, should develop its own strategy along the three axis of sustainable development as follows:

- **Social:**
  - Providing new methodologies and technologies to improve the livelihood of the local and global population.
- **Environment:**
  - Impacting positively the environment by developing new technologies for material processing, product development, water and

[40] p.13

air treatment, energy saving, reduction of global warming, *etc.*

- **Economic:**
  - Developing a new model of business for the shoe industries to re-design, commercialise and distribute new product, process and service solutions and therefore providing new jobs to local people.
  - Upgrading the industrial SME base by incorporating them into a “sustainable innovation” in the value chain, impacting also their growth and economic output.

Unfortunately, in only very rare cases the three sustainable development axes are taken into consideration when assessing the impact of a new technology. While many projects have been already dealing with the ecological problem from a process and material point of view (like for example the on-going European Commission sponsored CECmadeShoe research project), EUROShoE has contributed to sustainability because of the mass customization business model.

In fact, if we consider that a pair of shoes is manufactured only when the client has ordered them, then we can understand that no stock is held, no unsold shoes are returned, therefore reducing considerably the impact on the environment. If we look just to some simple calculations and, as was shown in previous studies (see for instance the outcomes of the analysis and surveys done in the EUROShoE project) that the market for customized shoes is around 30% of the global shoe market, then we can assume that more than 1.5 billion pair of shoes could be the target. According to some statistics the return on unsold shoes per year is around 20% and therefore the total number of returned shoes is more that 500 million pair (using conservative numbers). Now, according to a report published by Timberland Company [31], an already very sustainability conscious company, the energy to produce one pair of shoes is 3 kWh. Therefore the total energy necessary to produce these unsold shoes would be 1,500,000,000 kWh. This energy could be saved just by applying the mass customization business model.

If we add other environmental impact like the reduction of dangerous materials, transportation costs (and other energy saving), *etc.* it is easy to understand the large impact that a mass customization model may have on the environment. We are then considering a paradigm shift that originates from the idea of putting back the consumer at the centre of the producer attention, at the beginning and at the end of the whole life cycle of the shoe product. But, following the line of thought that we have just presented, we can also see how such consumer centric approach will eventually bring the consumer benefits on two separate levels: the individual one (a more comfortable and personalized shoe), and the collective one (a healthier environment in which only the energy needed to produce what will actually be purchased by a customer is consumed). This is another relevant argumentation in favour of mass customization.

6.4 Can Mass Customization be the Salvation of the Footwear Industry?

We now consider the most fundamental question: can a general widespread adoption of mass customization be the survival strategy for the footwear industry? Can it represent a salvation from the difficult recession phase it is going through?

Before stepping more decisively into these considerations, we think it worth recalling here some of the points highlighted in Chapter 2 in relation to which parts of the footwear world we are thinking of in presenting this competitive recipe. We have lengthily discussed there how the footwear world is, very schematically speaking, split into three groups of players that also identify well-defined geographical areas in the world:

- Those who are producing mainly for volumes and that are undergoing and almost unstoppable growth (the Far East and China in particular); the MC therapy is not foreseen for them. They don't need special therapies because there are not yet evident diseases to be cured. MC could be an interesting option for them, possibly as a special kind of manufacturing service requested by Western companies that are accustomed to working with their long distance supplier and now want extra level of service. But nothing more than that, at least in a short-term time scenario.

- Those who are mainly buying and distributing shoes serving consumers and satisfying their needs (north America and northern Europe) and, at the same time, rewarding their stakeholders with generous sale margins; MC can be an interesting option for this kind of large scale, massive retailers, as long as it is in response to customer demands. If consumers become more aware of the benefits in buying and wearing shoes that are produced and selected for their own feet and on the contribution that such choice can give to their health, shoe providers will have to modify their sales strategies towards mass customization. Then the problem will arise on how to set up the appropriate technological infrastructures and where to find the necessary production capacity. In these initial steps of the diffusion of the MC idea among shoe consumers, these aspects might not (yet) be problems, but in future such an evolution could cause a shortage of manufacturing resources and problems in relying on the appropriate supplies.

- Those who are mainly quality producers (value rather than volume) and who want to continue with such a strategy (mostly Latin America and southern Europe countries). For this last group of companies, mass customization can certainly represent one possible path to salvation and one (although not the only one) of the competitive strategies to be deployed.

So is mass customization the "salvation" of the footwear industry? Not certainly for the first group, which does not need to be salvaged; possibly for the second if consumers stimulate a move in that direction - without such consumer awareness, market players of this kind don't have a real reason to consider mass customization because they are not (or they don't feel they are) in such a desperate situation that would justify exceptional countermeasures.

So when we speak of salvation we specifically have in mind the third group of footwear companies: producing mid to high quality shoes, located in countries where labour costs are high, with their own manufacturing facilities that they don't want to loose (since they consider the manufacturing ability as a value to be preserved) and conscious that the only competition that they have a chance of winning will be on quality, services and added value. For these companies we believe mass customization can be at least one of the possible approaches to salvation. If at the time of the first theorization of the mass customization paradigm, there were been researchers who anticipated a future in which products of all kinds would have been made customized or customizable and mass production would have actually disappeared, these opinions have now changed.

None of the experts or researchers in this field is betting anymore on a future of that kind; more realistically they foresee scenarios in which lines of customized products live together with standard ones. What is likely to happen is that companies will adopt mixed strategies combining the traditional production of standard goods in those areas or market segments in which they feel they are strong and they can face the competition on quantities, with the offer of customized ones; the latter will qualify their market offer, will sell themselves for the added value they offer to consumers, and will, as we noted in Chapter 2, also fertilize and improve the traditional products.

Mass customization will more and more become one "common" option in the shoe business that companies can choose to increase their competitiveness; technology is proving that this is possible, consumers are learning to ask for it and are becoming aware of its advantages. Shoe companies will have to acknowledge this fact to take advantage of the market opportunity that it represents.

But also in this scenario, there is a time factor to be considered: who will be the fastest ones? The risk this question underlines is that there might some other market players in other parts of the world who are capable of intercepting this demand and who will then position themselves in this new emerging market segment.

To better explain what we mean with this comment on the "time factor", we would like to report about the announcement that was recently made on the Internet by an Hong Kong-based footwear trader that works in cooperation with shoe factories in China; the announcement was about "customized shoes" and

> *... a new production line in China which is solely for tailor-made shoes business.*

The message continued with a list of argumentations on why should such kind of shoes be made in China, because they are:

> *... Cost competitive. No one doubts that tailor-made shoes business is a labour -intensive business. In China, we can offer you our skilful shoe makers yet with much lower cost and much efficient production ...*

And on reasons for moving towards the production of customized shoes, because:

> *... Tailor-made shoe business could differentiate your company/ brand from your competitors ...*

And finally with a simple and neat explanation of how it works:

> *Simply provide us your customer's measurement, drawing, material and colour selection, then we can make the shoes for you. We can do the shipment on weekly or bi-weekly basis ...*

Mass customized shoes made easily and cheap for European companies that don't want to get involved with the complexity of this manufacturing operations and the burden of dedicated production units. So that part of the footwear world that apparently would not need MC for its survival is already monitoring and intercepting the trend, offering its services as it has done until now with traditional shoes. On the other hand, it is not a secret that the majority of the customized shoes that are currently being offered and sold to consumer are, in fact, produced in China (see the Nike, Adidas or Viavor cases). So the risk we see here is that if European companies in particular, do not move fast enough, someone else will decide to produce customized shoes for the entire world, as it has done so far for all the other kinds of shoes; and in such a case the regained competitiveness of the European footwear producers will very likely affect their brands (when they are there), their sales networks, their retailers; but the manufacturing ability will once again be lost.

To conclude, even though we cannot say that mass customization is yet a reality, there are signals that indicate that the investment, in terms of efforts and resources, by private entrepreneurs as well as by research institutions has not been in vain and that the day in which European consumers are given the possibility of buying shoes "made for their feet" is not so far away in time. However, this big achievement for consumers will not necessarily result in competitive leverage for European footwear makers if they don't move fast enough along the path of mass customization.

References

[1] Tseng M, Piller F, eds. (2003) The customer centric enterprise: advances in mass customization and personalization. Springer, New York

[2] Jovane F, Koren Y, Boër, CR (2003) Present and future of flexible automation: towards new paradigms. Keynote paper, 53rd general assembly of CIRP, Montreal, 24–30 August. CIRP Annals, 52(2): 543–560

[3] Piller F, Reichwald R, Möslein K (2000) Information as a critical success factor for mass customization, or: why even a customized shoe not always fits. Proceedings of the ASACIFSAM 2000 Conference, Montreal, July

[4] Piller F, Tseng M, eds. (2005) 2005 world congress on mass customization and personalization, Hong Kong, September

[5] Boër CR, Dulio S (2003) The Euro Shoe project: Europe's largest MCP research project. World congress on mass customization and personalization MCPC

[6] Carrie A (1999) Integrated clusters the future basis of competition. Int J Agile Management Systems 1(1): 45–50.

[7] Childe S (1998) The extended enterprise – a concept of cooperation. Production Planning & Control 9(4): 320–327

[8] Wiendhal H-P, Helms K, Lutz S (1999) Supply net - new methods for capacity flexibilisation in production networks. Proc 15th Int Conf computer-aided production engineering, pp. 609–610

[9] Bullinger HJ, Wagner F, Kürümlüoglu M (2003) Towards the extended user oriented shoe enterprise, the customer centric enterprise. Advances in Mass customization and Personalization, Springer, Berlin, pp. 451-463

[10] www.dell.com

[11] Dulio S, Boër CR (2004) Integrated Production Plant (IPP): an innovative laboratory for research projects in the footwear field. Int J Computer Integrated Manufacturing 17(7): 601 – 611

[12] Anderson DM (2004) Build-to-order and mass customization, the ultimate supply chain and lean manufacturing strategy for low-cost on-demand production without forecasts or fnventory. CIM Press

[13] Tu Q, Vonderembse MA, Ragu-Nathan TS (2001) The impact of time-based manufacturing practices on mass customization and value to customer. J Operations Management 19: 201-217

[14] Su JCP, Chang Y-L, Ferguson M (----) Evaluation of postponement structures to accomodate mass customization. J Operations Management 23: 305-318

[15] Barnett L, Rahimifard S, Newmann S (2004) Distributed scheduling to support mass customization in the shoe industry. J Computer Integrated Manufacturing 17(7): 623-632

[16] Fornasiero R, Zangiacomi A, Avai A (2004) Web cost simulation service for footwear sector. Int J Computer Integrated Manufacturing 17(7): 661-667

[17] Piller FT, Muller M (2004) A new marketing approach to mass customization. J Computer Integrated Manufacturing 17(7): 583-593
[18] Franke N, Piller F (2004) Toolkits for user innovation and design: exploring user interaction and value creation in the watch market. J Product Innovation Management 21
[19] Junior V, Stotko CM, Piller FT (2004) Spare parts on demands (sod) by leveraging mass customization manufacturing capabilities. Proc. 30th FISITA world automotive congress, Pedro de Esteban Altirriba (ed.), Barcelona, 23-27 May 2004, pp.17-27
[20] Kotha S (1995) Mass customization: implementing the paradigm for competitive advantage. Strategic Management J 16: 21-42
[21] Reichwald R, Piller FT, Jaeger S, Zanner S (2003) Economic evaluation of mini-plants for mass customization, the customer centric enterprise. Adv mass customization and personalization, Springer, Berlin, pp.51-69
[22] Boër CR, Dulio S, Guilizzoni M, Wu C (1999) A concept for an extended enterprise for shoe manufacturing technology in a framework for technical cooperation between Europe and China. Fourth CENIM Workshop
[23] Boër CR Jovane F (1996) Towards a new model of sustainable production: manufuturing. Annals CIRP 45(1): 415–420
[24] Zangiacomi A, Zhijian L, Sacco M, Boër CR (2004) Process planning and scheduling for mass customised shoe manufacturing. Int J Computer Integrated Manufacturing 17(7): 613 – 621
[25] Paris I, Handley D (2004) CAD usage and knowledge base technology in shoe design and development. Int J Computer Integrated Manufacturing 17(7):....
[26] Jatta F, Zanoni L, Fassi I, Negri S (2004) A roughing/cementing robotic cell for custom made shoe manufacture. Int J Computer Integrated Manufacturing 17(7):....
[27] Viganò G, Mottura S, Greci L, Sacco M, Boër CR (2004) Virtual reality as a support tool in the shoe life cycle. Int J Computer Integrated manufacturing 17(7): 653-660
[28] Fornasiero R, Zangiacomi A, Dulio S, Boër CR (2005) Evaluation of innovative business models for mass customization in the shoe sector. Third interdisciplinary world congress on mass customization and personalization, MCPC'05
[29] Redaelli C, Sacco M, Dulio S, Boër CR (2005) Analysis of cultural gap for customized product. Third Interdisciplinary world congress on mass customization and personalization, MCPC'05
[30] McDonough W, Braungart M, (2002) Cradle to cradle: remaking the way we make things, North Point Press
[31] The Timberland Company, Corporate Social Responsibility Report, 2005
[32] Ashford N, (2004) Major challenges to engineering education for sustainable development: what has to change to make it creative, effective, and acceptable to the established disciplines. Int J Sustainability Higher Education 3(4): 239–250
[33] Bruntland, G. (ed.), (1987) Our common future: The World Commission on Environment and Development. Oxford, Oxford University Press

Index

MIX
Papier aus verantwortungsvollen Quellen
Paper from responsible sources
FSC® C105338

If you have any concerns about our products,
you can contact us on
ProductSafety@springernature.com

In case Publisher is established outside the EU,
the EU authorized representative is:
Springer Nature Customer Service Center GmbH
Europaplatz 3, 69115 Heidelberg, Germany

Printed by Libri Plureos GmbH
in Hamburg, Germany